W9-AED-534

INFINITE DIMENSIONAL MORSE THEORY AND ITS APPLICATIONS

Notes du cours de Monsieur Kung-Ching Chang à la vingt-deuxième session du Séminaire de mathématiques supérieures / Séminaire scientifique OTAN (ASI 48/83), tenue au Département de mathématiques et de statistique de l'Université de Montréal du 27 juin au 15 juillet 1983. Cette session avait pour titre général « Méthodes topologiques en analyse non linéaire » et était placée sous les auspices de l'Organisation du Traité de l'Atlantique Nord, du ministère de l'Éducation du Québec, du Conseil de recherches en sciences naturelles et en génie du Canada et de l'Université de Montréal.

SÉMINAIRE DE MATHÉMATIQUES SUPÉRIEURES
SÉMINAIRE SCIENTIFIQUE OTAN (NATO ADVANCED STUDY INSTITUTE)
DÉPARTEMENT DE MATHÉMATIQUES ET DE STATISTIQUE — UNIVERSITÉ DE MONTRÉAL

INFINITE DIMENSIONAL MORSE THEORY AND ITS APPLICATIONS

KUNG-CHING CHANG
Peking University

1985
LES PRESSES DE L'UNIVERSITÉ DE MONTRÉAL
C.P. 6128, succ. «A», Montréal (Québec), Canada H3C 3J7

ISBN 2-7606-0734-8

DÉPÔT LÉGAL — 4e TRIMESTRE 1985 — BIBLIOTHÈQUE NATIONALE DU QUÉBEC

FOREWORD

These ten chapters contain the basic material in the lectures delivered at the 22nd session of the Séminaire de mathématiques supérieures at Montréal in 1983.

I would like to take this opportunity to express my appreciation of the invitation of Prof. A. Granas to give the lectures at the conference and to publish these notes in the collection.

I am particularly indebted to Prof. S.S. Chern for his invitation to the Mathematical Sciences Research Institute at Berkeley, California. In the favorable atmosphere, these notes were prepared.

I would also like to thank Mr. G. Tian and Mr. B. Chen for their valuable help in preparing these notes.

I am grateful to the editor of this book, Prof. Gert Sabidussi, for improving the English, and to Miss Johanne Beausoleil for her excellent typing.

Kung-Ching Chang

Mathematical Institute, Peking University, Beijing
July 30, 1984.

CONTENTS

INTRODUCTION

Degree theory has proved very useful in nonlinear analysis, in proving the existence of solutions and in estimating the number of solutions of an operator equation. The degree of an isolated zero of a vector field is assigned to be an integer describing the local behaviour of the vector field. In Morse theory, the zero of a gradient vector field, or the critical point of a function, is characterized by a series of homology groups. In this sense, Morse theory provides more fine structure than the degree in distinguishing the types of critical points. One may therefore hope to obtain more information from Morse theory in studying multiple solution problems in analysis.

However, the applications of the classical Morse theory to partial differential equations were very limited, because only nondegenerate functions, or nondegenerate critical points, were well studied, but the nondegeneracy of a solution of a partial differential equation occurring in practice is difficult to verify in advance, sometimes it is even not true. To extend the applications, there are two theories in front of us: (1) Isolated critical point theory due to Gromoll-Meyer [39], (2) Conley's Morse index theory on isolated invariant sets [29], [73]. Since the latter theory is much more general and abstract (even a variational structure is not needed), and since Conley's Morse index is a homotopy class, this theory is hard to read and the index is difficult to figure out. I would rather prefer to take an easier one: within a variational frame, we develop the Gromoll-Meyer theory in combining it with a result on the homotopy invariance which is an analogue of a theorem due to Conley. Fortunately, this theory is sufficient for many applications.

The aim of these lectures is to introduce the Morse theory systematically in this version, and to emphasize its applications in studying multiple solution problems in nonlinear analysis.

Sections 1 and 2 present the basic theory on Hilbert Riemannian manifolds. The two equivalent definitions of critical groups, the Morse lemma and its generalization — the splitting theorem, the Morse inequalities and a shifting theorem on isolated degenerate critical points, are studied.

Section 3 extends the classical Morse theory to Banach-Finsler manifolds. The nondegeneracy is defined, and the Morse inequalities are established.

The interplay between the Leray-Schauder index and the critical groups is studied in Section 4 as a local version of the Poincaré-Hopf theorem. In the same section, a formula connecting the Leray-Schauder degree with the relative Euler characteristic, and a minimax theorem generalizing the Ljusternik-Schnirelman category theorem via the cap product of a cohomology class with relative homology classes are obtained.

In Section 5 a three-critical-point theorem with applications is discussed.

Section 6 includes a homotopy invariance theorem of the critical groups, with an application to a bifurcation theorem due to Krasnoselskii and Rabinowitz.

Section 7 studies the homological characterization of the mountain pass point. According to this characterization a more general form of the mountain pass theorem is given, which covers new applications.

Sections 8-10 deal with applications to elliptic BVP and periodic solutions of Hamiltonian systems. The reader will find that there are many different and very interesting results in these sections. Some of them have been published in the literature, but the proofs given here are new and are presented in a unified way. Some of these results are published here for the first time. Section 8 contains some abstract theorems in the functional analytic framework. They will be used in Section 9 and Section 10. In Section 8, there is a finite dimensional reduction theory, which is a variant of the so-called saddle point reduction due to Amann [2]. The reader will find that the procedure of reduction has been considerably simplified. In Section 10, there is a simple proof of Arnold's conjecture on the number of fixed points of symplectic maps.

The lectures are not intended to be complete, neither as a systematic study of Morse theory, nor as concerns the presentation of a lot of applications. For instance, the equivariant theory [19], the nondegenerate critical manifold theory [18], Morse-Smale-Conley theory [74], [29], and the beautiful applications in the study of closed geodesics (for the vast literature the reader is referred to the book of Klingenberg [45]) as well as those in the study of gauge theory (cf. Atiyah-Bott [13]), are not dealt with in this work.

Section 0

PRELIMINARIES

The following notations and basic results will be used throughout the lectures.

Let M be a C^1 manifold, modelled on a Banach space, let f be a continuously differentiable function defined on M, which is denoted by $f \in C^1(M,\mathbb{R}^1)$.

$\forall p \in M$, $df(p)$ denotes the *differential* at p, $df(p) \in T_p(M)^*$, the cotangent space at p.

$\forall a \in \mathbb{R}^1$, $f_a = \{x \in M \mid f(x) \le a\}$ is called the *level set*. A point $p_o \in M$ is said to be a *critical point* of f if $df(p_o) = \theta$.

$K = \{p \in M \mid df(p) = \theta\}$ is called the *critical set* of f.

$\forall c \in \mathbb{R}^1$, K_c stands for $K \cap f^{-1}(c)$.

A real number c is said to be a *critical value* of f if $K_c \neq \emptyset$; otherwise, c is said to be a *regular value*.

A function $f \in C^1(M,\mathbb{R}^1)$ is said to satisfy the *Palais-Smale* (P.S. for abbreviation) *condition*, if for any sequence $\{p_n\} \subset M$,

$$\text{(the boundedness of } \{f(p_n)\}) \text{ plus } (df(p_n) \to \theta)$$

imply that

$\exists$ a convergent subsequence $p_{n'} \to p^*$ in M.

On the tangent bundle $T(M)$, if there is a function $\|.\|: T(M) \to \mathbb{R}^1_+$ such that

(1) $\|\cdot\|$ is continuous,

(2) $\forall p \in M$, let

$$\|\cdot\|_p = \|\cdot\| \big|_{T_p(M)},$$

then $\|\cdot\|_p$ is an equivalent norm on $T_p(M)$.

(3) $\forall p_0 \in M$, for each neighbourhood U of p_0 in M such that $T(U)$ is trivialized, and for each $k > 1$, there exists a neighbourhood V of p_0, such that $V \subset U$ and

$$\frac{1}{k}\|\cdot\|_p \le \|\cdot\|_{p_0} \le k\|\cdot\|_p \qquad \forall p \in V;$$

then $\|\cdot\|$ is called a *Finsler structure* on $T(M)$, and M is called a *Finsler manifold.*

One can show that for every paracompact Banach manifold M there is a Finsler structure such that M is a Finsler manifold.

Suppose that M is a Finsler manifold with Finsler structure $\|.\|$. One defines

$$\|x^*\| = \sup\{\langle x^*,x\rangle \mid \|x\|_p \le 1, \ x \in T_p(M)\}, \qquad \forall x^* \in T_p(M)^*,$$

where $\langle , \rangle$ is the duality between $T_p(M)^*$ and $T_p(M)$.

In particular, $\forall f \in C^1(M,\mathbb{R}^1)$, where M is a paracompact Banach manifold, $p \mapsto \|df(p)\|$ is continuous on M.

A vector field $\Xi \in T(M)$ is called a *pseudo-gradient vector field* (p.g.v.f. in short) on a subset $S \subset M$, if $\forall p \in S$,

$$\|\Xi(p)\| \le 2\|df(p)\|$$

and

$$\langle df(p), \Xi_{(p)} \rangle_p \geq \|df(p)\|^2,$$

where $\|X\|$ is the Finsler structure on $T(M)$, and $\langle , \rangle_p$ stands for the duality between $T_p(M)^*$ and $T_p(M)$.

For a function $f \in C^1(M, \mathbb{R}^1)$, it is proved (cf. Palais [61],[63]), that there exists a pseudo-gradient vector field defined on the set $\tilde{M} = M \backslash K$. By which one has the following deformation theorem:

(First deformation lemma) *Let* M *be a* C^2*-Finsler manifold. Suppose that* $f \in C^1(M, \mathbb{R}^1)$ *satisfies the P.S. condition. Assume that* c *is a real number, and that* N *is a closed neighbourhood of* K_c*. Then there is a continuous map* $\eta: [0,1] \times M \to M$, *as well as real numbers* $\bar{\varepsilon} > \varepsilon > 0$, *such that*

(1) $\eta(t,.)\big|_{Cf^{-1}[c-\bar{\varepsilon}, c+\bar{\varepsilon}]} = id\big|_{Cf^{-1}[c-\bar{\varepsilon}, c+\bar{\varepsilon}]}$,

(2) $\eta(0,.) = id$,

(3) $\eta(1, f_{c+\varepsilon} \backslash N) \subset f_{c-\varepsilon}$,

(4) $\forall t \in [0,1]$, $\eta(t,.): M \to M$ *is a homeomorphism.*

For a proof cf. Palais [61], Clark [28], and Rabinowitz [66].

In Morse theory, the *singular homology* and *singular cohomology theory* are applied. For a pair of topological spaces (X,Y) with $Y \subset X$, and an Abelian group G, $H_*(X,Y;G)$ stands for the *singular relative homology groups* with coefficient group G, and $H^*(X,Y;G)$ stands for the *relative cohomology ring*. For the definitions and basic properties, we refer the reader to Greenberg [38].

In case $Y = \emptyset$, we write $H_*(X,G) = H_*(X,\emptyset;G)$ and $H^*(X,G) = H^*(X,\emptyset;G)$. If, further, $X = M$ is a manifold,

$$\beta_q = \text{rank } H_q(M,G)$$

is called the qth *Betti number*, $q = 0,1,2,\ldots$ and

$$\chi(M) = \sum_{q=0}^{\infty} (-1)^q \beta_q$$

is called the *Euler characteristic* of the manifold M.

Now we turn to some notations and basic facts in the theory of partial differential equations.

Let $\Omega \subset \mathbb{R}^n$ be a bounded open domain with smooth boundary $\partial\Omega$. For a nonnegative integer vector $\alpha = (\alpha_1,\ldots,\alpha_n)$ we write

$$\partial^\alpha = \frac{\partial^{|\alpha|}}{\partial x_1^{\alpha_1}\ldots\partial x_n^{\alpha_n}}$$

to denote the differential operator, with $|\alpha| = \alpha_1 + \ldots + \alpha_n$.

Let $\mathcal{D}(\Omega)$ be the function space consisting of C^∞ functions with compact support in Ω, and let $\mathcal{D}'(\Omega)$ be the dual of $\mathcal{D}(\Omega)$, i.e., the Schwartz distribution space. For each integer $m \geq 0$, we denote

$$C^m(\overline{\Omega}) = \{u:\overline{\Omega} \to \mathbb{R}^1 \mid \partial^\alpha u \text{ is continuous on } \overline{\Omega},\ |\alpha| \leq m\},$$

with norm

$$\|u\|_m = \sum_{|\alpha|\leq m} \sup_{x\in\overline{\Omega}} |\partial^\alpha u(x)|.$$

For $p \geq 1$, and an integer $m \geq 0$, we denote

$$W_p^m(\Omega) = \{u \in L^p(\Omega) \mid \partial^\alpha u \in L^p(\Omega),\ |\alpha| \leq m\},$$

where L^p is the p-th power integrable Lebesgue space, and ∂^α is the differential operator in the distribution sense, with norm

$$\|u\|_{W_p^m} = \{\sum_{|\alpha|\leq m} \|\partial^\alpha u\|^p_{L^p(\Omega)}\}^{\frac{1}{p}}.$$

$W_p^m(\Omega)$ is called the *Sobolev space*. In particular, if $p = 2$, $H^m(\Omega)$ stands for $W_2^m(\Omega)$. The closure of $\mathcal{D}(\Omega)$ in the space $W_p^m(\Omega)$ ($H^m(\Omega)$ and $C^m(\bar{\Omega})$) is denoted by $\mathring{W}_p^m(\Omega)$ ($H_o^m(\Omega)$, $C_o^m(\bar{\Omega})$ respectively).

The dual space of $\mathring{W}_p^m(\Omega)$ (and $H_o^m(\Omega)$) is denoted by $W_{p'}^{-m}(\Omega)$ (and $H^{-m}(\Omega)$ resp.), where $\frac{1}{p} + \frac{1}{p'} = 1$.

The following inequalities are applied very frequently (cf. [37]):

<u>(Poincaré inequality)</u>

$$\left(\int_\Omega |u|^p \, dx\right)^{\frac{1}{p}} \leq C(\Omega)\left(\int_\Omega |\nabla u|^p \, dx\right)^{\frac{1}{p}} \qquad \forall u \in \mathring{W}_p^1(\Omega)$$

where ∇u denotes the gradient of u, and $C(\Omega)$ is a constant independent of u.

<u>(Sobolev inequality)</u>. Suppose that for $1 \leq p,r < \infty$ and integers $\ell \geq m \geq 0$, we have:

(1) If $\frac{1}{p} \leq \frac{1}{r} + \frac{\ell-m}{n}$, then the imbedding $W_p^\ell(\Omega) \hookrightarrow W_r^m(\Omega)$ is continuous. If the inequality $\leq$ is replaced by a strict inequality $<$, then the imbedding is compact.

(2) If $\frac{1}{p} \leq \frac{\ell-m}{n}$, then the imbedding $W_p^\ell(\Omega) \hookrightarrow C^m(\bar{\Omega})$ is continuous. If the inequality $\leq$ is replaced by a strict inequality $<$, then the imbedding is compact.

For a function $f \in C(\bar{\Omega}\times\mathbb{R}^1,\mathbb{R}^1)$, suppose that there exist constants α, $C > 0$ such that

$$|f(x,t)| \leq C(1+|t|^\alpha).$$

Then the following nonlinear operator:

$$u \mapsto f(x,u(x))$$

maps boundedly and continuously from $L^p(\Omega)$ to $L^q(\Omega)$, with $p = \alpha q$, for example, cf. [46]. The operator is called the *Nemytcki operator*. Applying this result, in combining with the Sobolev inequality, we see that the functional

$$J(u) = \int_\Omega f(x,u(x))\,dx$$

is well defined and continuous on the Sobolev space $H_0^1(\Omega)$ if

$$\alpha \le \frac{2n}{n-2}.$$

If, further, we assume that $f \in C^1(\bar\Omega\times\mathbb{R}^1,\mathbb{R}^1)$ satisfies the following growth condition:

$$|f_t'(x,t)| \le C(1+|t|^\alpha) \quad \text{with } \alpha \le \frac{n+2}{n-2},$$

then the functional J is C^1 on the space $H_0^1(\Omega)$, with differential

$$(dJ(u),v) = \int_\Omega f_t'(x,u(x))v(x)\,dx, \qquad \forall v \in H_0^1(\Omega).$$

Furthermore if $f \in C^2(\bar\Omega\times\mathbb{R}^1,\mathbb{R}^1)$, satisfies

$$|f_{tt}''(x,t)| \le C(1+|t|^\alpha) \quad \text{with } \alpha \le \frac{4}{n-2},$$

then J is C^2 on $H_0^1(\Omega)$, with

$$d^2J(u)(v,w) = \int_\Omega f_{tt}''(x,u(x))v(x)w(x)\,dx \qquad \forall v,w \in H_0^1(\Omega).$$

As to the Laplacian $-\Delta$ defined on $L^2(\Omega)$, with domain $D(-\Delta) = H^2(\Omega) \cap H_0^1(\Omega)$, it is a self-adjoint operator. The operator $-\Delta$ can be extended from $W_p^2 \cap \mathring{W}_p^1(\Omega)$ to $L^p(\Omega)$, $1 < p < \infty$, continuously.

It is known that $\ker(-\Delta) = \{\theta\}$, and $K = (-\Delta)^{-1}$ maps $L^p(\Omega)$ into itself continuously, and is a compact operator. Also the operator K maps $L^2(\Omega)$ into $H_0^1(\Omega)$, such that

$$\int_\Omega u\cdot v\,dx = (Ku,v)_{H_0^1} \qquad \forall v \in H_0^1(\Omega), \quad u \in L^2(\Omega).$$

The following boundary value problem will be considered:

$$(0.1)\qquad \begin{aligned} -\Delta u &= g(x,u(x)) \quad \text{in } \Omega \\ u|_{\partial\Omega} &= 0 \end{aligned}$$

where the following growth condition on $g \in C(\bar{\Omega}\times\mathbb{R}^1,\mathbb{R}^1)$ is assumed:

$$|g(x,t)| \leq C(1+|t|^{\alpha}) \qquad \alpha < \frac{n+2}{n-2},$$

for constants $C, \ \alpha > 0$. We say that $u_0 \in H_0^1(\Omega)$ is a *weak solution* of (0.1) if

$$\int_\Omega [\nabla u_0 \cdot \nabla v - g(x,u_0(x))v(x)]\, dx = 0 \qquad \forall v \in H_0^1(\Omega).$$

If we define the functional

$$J(u) = \int_\Omega (\tfrac{1}{2}|\nabla u|^2 - G(x,u(x)))dx,$$

where

$$G(x,t) = \int_0^t g(x,\xi)\, d\xi;$$

then J is C^1 on $H_0^1(\Omega)$ with

$$(dJ(u),v) = \int_\Omega [\nabla u\cdot\nabla v - g(x,u(x))\cdot v(x)]\, dx \qquad \forall v \in H_0^1(\Omega).$$

This means that the differential equation (0.1) is just the *Euler equation* of the functional J and the weak solutions of (0.1) are critical points of J, and conversely.

Since it is well known that weak solutions of (0.1) are classical solutions of (0.1), if the function g is smooth enough (cf. [37]), we are satisfied to look for weak solutions of (0.1), i.e., the critical points of J.

Section 1

CRITICAL GROUPS AND THE MORSE TYPE NUMBERS

The basic Morse theory is set up by two steps:

(1) Locally, we attach a sequence of groups, which we call critical groups, in describing the local behaviour of a function f near its critical point.

(2) Globally, a sequence of numbers, which count the critical points in accordance with the critical groups, is defined. These numbers are called Morse type numbers. We study the relationship between these Morse type numbers and the topological invariants of the underlying manifold via the deformation property.

DEFINITION 1.1. Let p be an isolated critical point of f, and let $c = f(p)$. We call

$$C_q(f,p) = H_q(f_c \cap U_p, (f_c \setminus \{p\}) \cap U_p; G)$$

the qth *critical group*, with coefficient group G of f at p, $q = 0,1,2,\ldots$; where U_p is a neighbourhood of p such that $K \cap (f_c \cap U_p) = \{p\}$, and $H_*(X,Y;G)$ stands for the singular relative homology groups with Ablian coefficient group G.

According to the excision property of the singular relative homology groups, the critical groups are well defined; i.e., they do not depend upon the special choice of the neighbourhood U_p. Now we give some examples:

EXAMPLE 1. Let p be an isolated local minimum of a function $f \in C(M,\mathbb{R}^1)$, then

$$C_q(f,p) = \begin{cases} G & q = 0 \\ 0 & q \neq 0. \end{cases}$$

In a finite dimensional manifold M^n, if p is an isolated local maximum of a function $f \in C(M^n,\mathbb{R}^1)$, then

$$C_q(f,p) = \begin{cases} G & q = n \\ 0 & q \neq n. \end{cases}$$

EXAMPLE 2. Let M be a 1-dimensional manifold, if p is an isolated critical point of a function $f \in C(M,\mathbb{R}^1)$ which is neither a local maximum nor a local minimum, then we have

$$C_0(f,p) = C_1(f,p) = 0.$$

EXAMPLE 3 (Monkey saddle). Let f be the function $x^3 - 3xy^2$ defined on $\mathbb{R}^2$, then we have

$$C_0(f,\theta) = C_2(f,\theta) = 0, \text{ and } C_1(f,\theta) = G \oplus G.$$

Suppose now that M is a Hilbert-Riemannian manifold, $f \in C^2(M,\mathbb{R}^1)$, $p \in K$ is called *nondegenerate*, if the Hessian $d^2f(p)$ has a bounded inverse. Since $A = d^2f(p)$ is a self-adjoint operator which possesses a resolution of identity, we call the dimension of the negative space corresponding to the spectral resolution, the *Morse index* of p, denoted by $\mathrm{ind}(f,p)$. Now we are in a position to compute the critical groups of a nondegenerate critical point via its Morse index.

We appeal to the following Morse lemma, but postpone the proof to the next section, in which a general splitting lemma will be proved.

(Morse lemma). *Suppose that* $f \in C^2(M,\mathbb{R}^1)$, *and that* p *is a nondegenerate critical point; then there exists a neighbourhood* U_p *of* p *and a local diffeomorphism* $\Phi : U_p \to T_p(M)$ *with* $\Phi(p) = \theta$, *such that*

$$f \circ \Phi^{-1}(\xi) = f(p) + \tfrac{1}{2}(d^2f(p)\xi,\xi) \qquad \forall \xi \in \Phi(U_p)$$

where $(\ ,\ )$ *is the inner product of the Hilbert space* H *on which the Riemannian manifold* M *is modelled.*

The following theorem relates the critical groups of a nondegenerate critical point with its Morse index.

THEOREM 1.1. *Suppose that* $f \in C^2(M,\mathbb{R}^1)$, *and* p *is a nondegenerate critical point of* f *with Morse index* j; *then*

$$C_q(f,p) = \begin{cases} G & q = j \\ 0 & q \neq j. \end{cases} \tag{1.1}$$

PROOF. According to the Morse lemma, we may restrict ourselves to a special case where f is a quadratic function on the Hilbert space H: $f(x) = \frac{1}{2}(Ax,x)$, where A is a bounded, invertible, self-adjoint operator. Let $P_\pm$ be the orthogonal projection onto the positive (negative) subspace $H_\pm$ with respect to the spectral decomposition of A. We have

$$f(x) = \tfrac{1}{2}(\|(AP_+)^{\frac{1}{2}}x\|^2 - \|(-AP_-)^{\frac{1}{2}}x\|^2).$$

On an ε-ball B_ε centered at θ, it is easily seen that

$$B_\varepsilon \cap f_0 = \{x \in H \mid \|x\| \le \varepsilon,\ \|y_+\| \le \|y_-\| \text{ where } y_\pm = (\pm AP_\pm)^{\frac{1}{2}}x\}.$$

We define a deformation:

$$\eta(t,x) = y_- + ty_+ \qquad (t,x) \in [0,1] \times (B_\varepsilon \cap f_0);$$

then it is a strong deformation retract from $(B_\varepsilon \cap f_0,\ B_\varepsilon \cap (f_0\backslash\{\theta\})$ to $(H_- \cap B_\varepsilon,\ (H_-\backslash\{\theta\}) \cap B_\varepsilon)$. Thus

$$C_q(f,p) \cong H_q(f_o \cap B_\varepsilon, (f_o \backslash \{\theta\}) \cap B_\varepsilon; G)$$
$$\cong H_q(H_- \cap B_\varepsilon, (H_- \backslash \{\theta\}) \cap B_\varepsilon; G)$$
$$\cong H_q(B^j, S^{j-1}) \cong \begin{cases} G & q = j \\ 0 & q \neq j \end{cases}$$

if j is finite. Nevertheless, for $j = +\infty$, since S^∞ is contractible, we always have $C_*(f,p) \cong 0$. The conclusion is proved. □

In the following, we assume that f has only isolated critical values, each of them corresponds to finite number critical points; say

$$\dots < c_{-2} < c_{-1} < c_o < c_1 < c_2 < \dots,$$

are critical values with

$$K_{c_i} = \{z_j^i\}_{j=1}^{m_i}, \qquad i = 0, \pm 1, \pm 2, \dots .$$

We choose

$$0 < \varepsilon_i < \text{Min}\{c_{i+1} - c_i, c_i - c_{i-1}\}, \qquad i = 0, \pm 1, \pm 2, \dots .$$

DEFINITION 1.2. For a pair of regular values $a < b$, we call

$$M_q(a,b) = \sum_{a < c_i < b} \text{rank } H_q(f_{c_i + \varepsilon_i}, f_{c_i - \varepsilon_i}; G),$$

the qth *Morse type number* of the function with respect to (a,b), $q = 0,1,2,\dots$

For functions f satisfying the P.S. condition, the Morse type numbers $M_q(a,b)$, $q = 0,1,2,\dots$, are well defined according to the first deformation lemma, i.e., they are independent of the special choice of $\{\varepsilon_i\}$.

Now, we shall obtain a relationship between the Morse type numbers and the critical groups of critical points.

The following deformation lemma is crucial in connecting different level sets.

A function f defined on M, is said of the *class* C^{2-0}, if f is C^1, and df is locally Lipschitzian.

THEOREM 1.2 (Second deformation lemma). *Suppose that* $f \in C^{2-0}(M,\mathbb{R}^1)$ *satisfies the P.S. condition, and that* c *is the only critical value of* f *in the interval* $[a,b) \subset \mathbb{R}^1$. *Assume that* c *corresponds to at most a finite number of critical points. Then* f_c *is a strong deformation retract of* f_b, *i.e., there exists a continuous* $\tau:[0,1] \times (f_b\backslash K_b) \to f_b\backslash K_b$ *such that*

$$\begin{cases} \tau(0,.) = \text{id}, \\ \tau(t,.)|_{f_c} = \text{id}|_{f_c}, \\ \tau(1,f_b\backslash K_b) \subset f_c. \end{cases}$$

PROOF. 1° For each $x_o \in f_o\backslash(f_c \cup K_b)$, we define a flow as follows

$$\begin{cases} \dot{\sigma}(t,x_o) = \dfrac{-(f(x_o)-c)V(\sigma(t,x_o))}{\langle df(\sigma(t,x_o)),V(\sigma(t,x_o))\rangle} \\ \sigma(0,x_o) = x_o \end{cases}$$

where V is the pseudo-gradient vector field of f over $\tilde{M} = M\backslash K$. It is easily seen that

$$\frac{d}{dt} f(\sigma(t,x_o)) = -(f(x_o)-c),$$

so that

$$f(\sigma(t,x_o)) = (1-t)f(x_o) + ct.$$

This implies that $\sigma(t,x_o)$ is defined on $[0,1) \times (f_b\backslash(f_c \cup K_b)$ and

$$\lim_{t\to 1-0} f(\sigma(t,x_o)) = 0.$$

However, the denominator of the right hand side of the ODE approaches 0 as σ

approaches a critical point, so that we cannot say at this moment that the solution $\sigma(t,x_o)$ is extentable to $t = 1$.

2° We shall prove that the limit $\lim_{t\to 1-0} \sigma(t,x_o)$ does exist. Let $K_c = \{Z_i\}_{i=1}^N$; then either one of the following cases occurs:

(a) $\inf_{t\in[0,1)} \operatorname{dist}(\sigma(t,x_o),Z_i) > 0 \quad \forall i = 1,\dots,N.$

(b) $\inf_{t\in[0,1)} \operatorname{dist}(\sigma(t,x_o),Z_i) = 0 \quad$ for some i.

In case (a), there exists an $\alpha > 0$ such that

$$\inf_{t\in[0,1)} \|df(\sigma(t,x_o)\| \geq \alpha$$

so that

$$\begin{aligned} \operatorname{dist}(\sigma(t_2,x_o),\sigma(t_1,x_o)) &\leq \int_{t_1}^{t_2} \|\tfrac{d\sigma}{dt}\| \, dt \\ &\leq (f(x_o)-c) \int_{t_1}^{t_2} \frac{\|V(\sigma)\|}{<df(\sigma),V(\sigma)>}\Big|_{\sigma=\sigma(t,x_o)} dt \\ &\leq \frac{2}{\alpha} (f(x_o-c)(t_2-t_1). \end{aligned}$$

This implies that the limit $\lim_{t\to 1-0} \sigma(t,x_o)$ does exist.

In case (b), we shall prove

$$\lim_{t\to 1-0} \sigma(t,x_o) = Z_i.$$

If not, there exists $\varepsilon > 0$ such that for an infinite sequence of disjoint intervals $[t_j,t_j^*] \subset [0,1)$ one has $\sigma(t,x_o) \in \overline{B}(Z_i,2\varepsilon)\backslash\mathring{B}(Z_i,\varepsilon)$, for $t \in [t_j,t_j^*]$, $j = 1,2,\dots$, where $B(Z,\varepsilon)$ is a ball centered at Z with radius ε. The $\varepsilon > 0$ is so chosen that

$$\alpha = \inf_{t\in[t_j,t_j^*]} \|df(\sigma(t,x_o))\| > 0.$$

Then we get a contradiction:

$$\varepsilon \le \text{dist}(\sigma(t_j^*,x_o),\sigma(t_j,x_o))$$

$$\le \int_{t_j}^{t_j^*} \left\|\frac{d\sigma}{dt}\right\| dt$$

$$\le \frac{2}{\alpha}\,(f(x_o-c)(t_j^*-t_j) \to 0.$$

3° Now we define a deformation retract as follows:

$$\tau(t,x_o) = \begin{cases} \sigma(t,x_o) & \text{if } (t,x_o) \in [0,1) \times (f_b\backslash f_c) \\ \lim\limits_{t\to 1-0} \sigma(t,x_o) & \text{if } (t,x_o) \in \{1\} \times (f_b\backslash f_c) \\ x_o & \text{if } (t,x_o) \in [0,1] \times f_c. \end{cases}$$

One has to verify the continuity of τ . Four cases are distinguished:

(a) $(t,x) \in [0,1] \times \mathring{f}_c$,

(b) $(t,x) \in [0,1) \times (f_b\backslash f_c)$,

(c) $(t,x) \in \{1\} \times (f_b\backslash f_c)$,

(d) $(t,x) \in [0,1] \times f^{-1}(c)$.

Case (a) is trivial, case (b) is verified by the fundamental theorem of ODE.

Verification for the case (c). Let $x^* = \lim\limits_{t\to 1-0} \sigma(t,x_o)$. We only want to prove the continuity of τ at those x_o where x^* is a critical point of f. We shall prove: $\forall\varepsilon > 0$, $\exists\delta > 0$ such that $\text{dist}(x_o,y) < \delta$ and $1-\delta < t$ imply

$$\text{dist}(\tau(t,y),x^*) < \varepsilon,$$

where we may assume without loss of generality that $y \notin f^{-1}(c)$.

Choosing suitable $\delta_1 > 0$ and $\delta_2 = \delta_2(\delta_1) > 0$ such that

$$\text{dist}(\tau(1-\delta_1,y),x^*) < \frac{\varepsilon}{2} \quad \text{for } \text{dist}(y,x_o) < \delta_2.$$

This is available, because we have $\delta_1 > 0$ such that

$$\mathrm{dist}(\tau(t,x_o),x^*) < \frac{\varepsilon}{4} \quad \text{for } t \in [1-\delta_1,1)$$

and then we have $\delta_2 > 0$ such that

$$\mathrm{dist}(\tau(t,x_o),\tau(t,y)) < \frac{\varepsilon}{4} \quad \text{for } \mathrm{dist}(y,x_o) < \delta_2, \quad t \in [0,1-\delta_1].$$

These imply that

$$\mathrm{dist}(\tau(1-\delta_1,y),x^*) < \frac{\varepsilon}{2} .$$

Without loss of generality we may assume that no critical points are in $\overline{B}(x^*,\varepsilon)\setminus B(x^*,\varepsilon/2)$. Thus

$$\alpha^* = \inf_{x\in\overline{B}(x^*,\varepsilon)\setminus B(x^*,\varepsilon/2)} \|df(x)\| > 0.$$

Determine $\delta_1 > 0$ so small that

$$(b-c)\delta_1 < \tfrac{1}{4}\varepsilon\alpha^*.$$

Then we shall prove that

$$\mathrm{dist}(\tau(t,y),x^*) < \varepsilon \qquad \forall (t,y) \in [1-\delta_1,1) \times B(x_o,\delta_2).$$

Because if not, $\exists t' < t''$ and $y_o \in B(x_o,\delta_2)$ such that

$$\frac{\varepsilon}{2} \le \mathrm{dist}(\tau(t,y_o),x^*) \le \varepsilon \qquad \text{for } t \in [t',t''] \subset [1-\delta_1,1)$$

with

$$\mathrm{dist}(\tau(t',y_o),x^*) = \frac{\varepsilon}{2}, \quad \mathrm{dist}(\tau(t'',y_o),x^*) = \varepsilon.$$

Then

$$\begin{aligned}\frac{\varepsilon}{2} &\le \mathrm{dist}(\tau(t'',y_o),\tau(t',y_o)) \\ &\le \int_{t'}^{t''} \left\|\frac{d\sigma}{dt}\right\| dt \\ &\le \frac{2}{\alpha^*}(b-c)\delta_1 < \frac{\varepsilon}{2} .\end{aligned}$$

This is a contradiction.

Case (d) is verified in the same way. Since now $x^* = x_o$, y_o is initially chosen in the ball $B(x_o,\varepsilon/2)$, only the last part of the above proof is applied. □

REMARK 1.1. The second deformation lemma was given by E. Rothe [70] in case M = Hilbert space, see also Chang [22].

The P.S. condition of the function f is frequently employed via deformation, in critical point theory. The following definition is used in some sense to replace the P.S. condition.

DEFINITION 1.3. A function $f \in C^1(M,\mathbb{R}^1)$ is said to possess the *deformation property*, if for each pair $a < b$,

(1) $K \cap f^{-1}(a,b) = \emptyset$ implies that f_a is a strong deformation retract of $f_b \backslash K_b$;

(2) $f(K)$ is a closed subset.

Thus, a function $f \in C^{2-0}(M,\mathbb{R}^1)$ with isolated K, satisfying the P.S. condition possesses the deformation property.

According to the deformation property, we obtain the following

THEOREM 1.3. *Assume that the function* $f \in C^1(M,\mathbb{R}^1)$ *possesses the deformation property, and that* c *is an isolated critical value of* f. *Let* $\varepsilon > 0$ *be so small such that* c *is the unique critical value in the inverval* $[c-\varepsilon,c+\varepsilon]$. *Then we have*

$$H_*(f_{c+\varepsilon},f_{c-\varepsilon};G) \cong H_*(f_c,f_c\backslash K_c;G) \cong \bigoplus_{j=1}^{m} C_*(f,z_j)$$

where $\{z_j\}_1^m = K_c$.

PROOF. By the deformation property and the homotopy invariance of the singular homology groups, we have

$$H_q(f_{c+\varepsilon},f_{c-\varepsilon};G) \cong H_q(f_c,f_{c-\varepsilon};G),$$

and

$$H_q(f_c\backslash K_c,f_{c-\varepsilon};G) \cong H_q(f_{c-\varepsilon},f_{c-\varepsilon};G) \cong 0.$$

Applying the exactness of the singular homology groups to the triple $(f_c,f_c\backslash K_c,f_{c-\varepsilon})$:

$$\dots \to H_q(f_c\backslash K_c,f_{c-\varepsilon}) \to H_q(f_c,f_{c-\varepsilon}) \to H_q(f_c,f_c\backslash K_c) \to H_{q-1}(f_c\backslash K_c,f_{c-\varepsilon}) \to \dots,$$

we find

$$0 \to H_q(f_c,f_{c-\varepsilon}) \to H_q(f_c,f_c\backslash K_c) \to 0,$$

that is

$$H_q(f_c,f_{c-\varepsilon}) \cong H_q(f_c,f_c\backslash K_c). \tag{1.2}$$

The first isomorphism is proved.

Furthermore, by using the excision property of singular homology groups, we may decompose the relative singular homology groups into critical groups:

$$\begin{aligned} H_*(f_{c+\varepsilon},f_{c-\varepsilon};G) &\cong H_*(f_c,f_c\backslash K_c;G) \\ &\cong H_*(f_c \cap \bigcup_{j=1}^{m} B(z_j,\varepsilon),f_c \cap \bigcup_{j=1}^{m} (B(z_j,\varepsilon)\backslash\{z_j\});G) \qquad (1.3) \\ &\cong \bigoplus_{j=1}^{m} C_*(f,z_j) \end{aligned}$$

for $\varepsilon > 0$ small enough, where $B(x,\varepsilon)$ is the ball centered at x with radius ε. The proof is finished. □

COROLLARY 1.1. $M_q(a,b) = \sum_{a<c_i<b} \sum_{j=1}^{m_i} \operatorname{rank} C_q(f,z_j^i)$, $q = 0,1,2,\dots$.

This is the connection between the critical groups and the Morse type numbers.

Now we turn to investigate the relationship between these Morse type numbers and the topological invariants of the underlying manifold, or level sets of f. Let us recall that for an exact sequence of groups

$$\cdots \xrightarrow{i} A \xrightarrow{j} B \longrightarrow \cdots$$

we have the following basic fact

$$\text{rank } A = \text{rank Im}(i) + \text{rank Im}(j).$$

Applying it to the following exact sequence:

$$\cdots \to H_q(Y,Z) \to H_q(X,Z) \to H_q(X,Y) \to H_{q-1}(Y,Z) \to \cdots \to 0$$

for a triple (X,Y,Z), we obtain the subadditivity of the following functions:

$$S_q(X,Y) = \sum_{j=0}^{q} (-1)^j \text{ rank } H_j(X,Y)$$

under the assumption that all $H_j(X,Y)$ have finite ranks, as well as the additivity of the Euler characteristic

$$\chi(X,Y) = \sum_{j=0}^{\infty} (-1)^j \text{ rank } H_j(X,Y)$$

if further, it is convergent; i.e., we have

(1.4) $$S_q(X,Z) \le S_q(X,Y) + S_q(Y,Z), \qquad q = 0,1,2,\ldots$$

and

(1.5) $$\chi(X,Z) = \chi(X,Y) + \chi(Y,Z).$$

A repetition of (1.4) yields the *Morse inequality*. Namely:

THEOREM 1.4. *Let* f *be a* C^1 *function possessing the deformation property on* M. *Assume that* f *has only isolated critical values, each of them corresponding to a finite number of critical points, with finite rank critical groups. Suppose that* $a < b$ *are regular values of* f. *Then we have the following inequa-*

lities:

$$(1.6)\quad M_q - M_{q-1} + \dots + (-1)^q M_o \geq \beta_q - \beta_{q-1} + \dots + (-1)^q \beta_o, \qquad q = 0,1,2,\dots,$$

where $M_q = M_q(a,b)$ *is the Morse type number w.r. to* (a,b), *and*

$$\beta_q = \beta_q(a,b) = \text{rank } H_q(f_b, f_a; G), \qquad q = 0,1,2,\dots .$$

Furthermore, we have

$$(1.7)\qquad \sum_{q=0}^{\infty} (-1)^q M_q = \sum_{q=0}^{\infty} (-1)^q \beta_q$$

if the left hand side converges.

Now we are able to give a geometrical interpretation of these Morse inequalities. Let f be a nondegenerate function (i.e., all critical points are nondegenerate), defined on a compact smooth manifold M. If we choose $b > \max_{x \in M} f(x)$, and $a < \min_{x \in M} f(x)$, then the Morse type number M_q is interpreted to be the number of critical points with Morse index q, and the number β_q is just the qth Betti number of the manifold M, $q = 0,1,\dots$. The LHS of (1.6) are alternating summations of these critical point numbers, and the RHS of (1.6) are alternating summations of the Betti numbers.

REMARK 1.2. The same inequalities (1.6) and identity (1.7) hold if M_q and β_q are replaced by their singular cohomological counterparts:

$$M^q = M^q(a,b) = \sum_{a<c_i<b} \sum_{j=1}^{m_i} \text{rank } C^q(f, z_j^i),$$

$$\beta^q = \beta^q(a,b) = \text{rank } H^q(f_b, f_a; G),$$

where $C^q(f,p) = H^q(f_c \cap U_p, (f_c \setminus \{p\}) \cap U_p)$, with $c = f(p)$, and U_p is a neighbourhood of p in which p is the unique critical point.

REMARK 1.3. Now we extend Theorem 1.4 to functions defined on a manifold M with boundary ∂M. The same proof works and the same conclusion holds for functions f under the following assumptions:

(1) $K \cap \partial M = \emptyset$;

(2) There exists a p.g.v.f. V of f such that $-V\big|_{\partial M\backslash f^{-1}(a)}$ directs inward (i.e., the negative pseudo-gradient flow $\eta(t,x) \in \mathring{M}$ $\forall x \in \partial M\backslash f^{-1}(a)$, $\forall t > 0$ small).

REMARK 1.4. The Morse inequalities (1.6) can be written in a more compact form. Define two formal series:

$$P(a,b;t) = \sum t^k\beta_k(a,b), \quad M(a,b;t) = \sum t^k M_k(a,b).$$

Then (1.6) is equivalent to saying that there exists $Q(t) = \sum q_k t^k$ with $q_k \geq 0$ such that $M(a,b;t) - P(a,b;t) = (1+t)Q(t)$ (cf. Bott [19]).

REMARK 1.5. The material of this section is based on the work due to E. Rothe [71].

Now we turn to define critical groups for an isolated critical set. A subset $S \subset K$ is called an *isolated critical set* of the function $f \in C^1(M,\mathbb{R}^1)$ satisfying the P.S. condition, if there exist an open set $\mathcal{O}$ and an interval $(\alpha,\beta) \subset \mathbb{R}^1$ such that

$$S = K \cap f^{-1}(\alpha,\beta) \cap \tilde{\mathcal{O}} \text{ and } K \cap (f^{-1}(\alpha) \cup f^{-1}(\beta)) \cap \tilde{\mathcal{O}} = \emptyset,$$

where $\tilde{\mathcal{O}} = \bigcup_{-\infty<t<+\infty} \eta(t,\mathcal{O})$, and η is the negative gradient flow.

One easily verifies that η can be replaced by a negative pseudo-gradient flow.

For an isolated critical set, let us denote

$$[S] = \{x \in M \mid \eta(\overline{\mathbb{R}}^1_+,x) \cap S \neq \emptyset\},$$

where $\eta(\overline{\mathbb{R}}^1_+,x) = cl\{\eta(t,x) \mid t \in \mathbb{R}^1_+\}$. $[S]$ is called the *invariant hull of* S, which depends on the p.g.v.f. by definition.

Similar to Theorem 1.3 we have the following

THEOREM 1.5. $H_*(f_\beta \cap \tilde{O}, f_\alpha \cap \tilde{O}; G) \cong H_*(f_\beta \cap \tilde{O}, (f_\beta \setminus [S]) \cap \tilde{O}; G)$.

The tubular set $\tilde{O}$ depends on the pseudo-gradient vector field V, it is denoted by $\tilde{O}_V$. It is easy to verify that the set $\tilde{O}_V$ can be deformed to the set $\tilde{O}_{df}$. Thus the homology groups $H_*(f_\beta \cap \tilde{O}, (f_\beta \setminus [S]) \cap \tilde{O}; G)$ depend on the critical set S and the function f only, but are independent of the special choice of the p.g.v.f. We define them to be the *critical groups of the isolated critical set* S.

REMARK 1.6. In case M is compact, the critical groups of an isolated critical set S, in some sense, are an analogy of the Conley index of an isolated invariant set of the vector field df (cf. Conley [29]). It will be discussed in Section 2.

At the end of this section, we improve Example 2 as follows.

EXAMPLE 4. If $f \in C^{2-0}(M^n, \mathbf{R}^1)$, and if p is an isolated critical point of f, which is neither a local maximum nor a local minimum, then

$$C_o(f,p) = C_n(f,p) = 0.$$

PROOF. We choose a path-connected neighbourhood U of p such that p is the only critical point of f in U, and such that $f_{c+\varepsilon} \cap U = U$ for some $\varepsilon > 0$, where $c = f(p)$. Let $U_+ = \bigcup_{t \geq 0} \eta(t,U)$, where η is the negative gradient flow of f. Then we have

$$C_*(f,p) \cong H_*(f_c \cap U_+, (f_c \setminus \{p\}) \cap U_+) \cong H_*(f_{c+\varepsilon} \cap U_+, (f_c \setminus \{p\}) \cap U_+).$$

Since $f_{c+\varepsilon} \cap U_+$ is path-connected, and $(f_c \setminus \{p\}) \cap U_+$ is non-empty, we obtain $C_o(f,p) = 0$. Similarly we prove $C_n(f,p) = 0$.

Section 2

GROMOLL-MEYER THEORY

The contributions of Gromoll and Meyer to isolated critical point theory are threefold: (1) A splitting lemma, which is a generalization of the Morse lemma (cf. the previous section). (2) An alternative definition of the critical groups. (3) A shifting theorem which reduces the critical groups of a degenerate isolated point to critical groups of the function restricted to its degenerate submanifold. In this section, we shall rewrite their theory with slight improvements, and prove the equivalence of the two definitions of critical groups.

THEOREM 2.1. *Suppose that* U *is a neighbourhood of* θ *in a Hilbert space* H, *and that* $f \in C^2(U,\mathbb{R}^1)$. *Assume that* θ *is the only critical point of* f, *and that* $A = d^2f(\theta)$ *with kernel* N. *If* 0 *is either an isolated point of the spectrum* $\sigma(A)$ *or does not belong to* $\sigma(A)$, *then there exists a ball* B_δ, $\delta > 0$, *centered at* θ, *an origin-preserving local homeomorphism* Φ *defined on* B_δ, *and a* C^1 *mapping* $h: B_\delta \cap N \to N^\perp$ *such that*

$$f \circ \Phi(z+y) = \tfrac{1}{2}(Az,z) + f(h(y)+y), \qquad \forall x \in B_\delta, \tag{2.1}$$

where $y = P_N x$, $z = P_{N^\perp}x$, *and* P_N *is the orthogonal projection onto the subspace* N.

PROOF. 1° Decomposing the space H into $N \oplus N^\perp$, we have

$$d_z f(\theta_1+\theta_2) = \theta_1 \qquad (\theta_1 = P_{N^\perp}\theta, \theta_2 = P_N\theta)$$

$$d_z^2 f(\theta_1+\theta_2) = A|_{N^\perp}.$$

On account of the implicit function theorem, there is a function $h:B_\delta \cap N \to N^\perp$, $\delta > 0$, such that

$$d_z f(y+h(y)) = \theta_1.$$

Let $u = z - h(y)$, and let

(2.2) $$F(u,y) = f(z+y) - f(h(y)+y).$$

(2.3) $$F_2(u) = \tfrac{1}{2}(Au,u).$$

Then we obtain

$$F(\theta_1,y) = 0,$$

$$d_u F(\theta_1,y) = d_z f(h(y)+y) = \theta_1,$$

$$d_u^2 F(\theta_1,\theta_2) = d_z^2 f(\theta) = A|_{N^\perp}.$$

2° Define $\xi:(u,y) \mapsto u_o \in F_2^{-1} \circ F(u,y) \cap \{\eta(t,u) \mid |t| < \|u\|\}$, where η is the flow defined by the following ODE:

$$\dot\eta(s) = -\frac{A\eta(s)}{\|A\eta(s)\|}$$

$$\eta(0) = u.$$

CLAIM. η *is well defined for* $|t| < \|u\|$. Since

$$\|\eta(s)-u\| \le |s|$$

we have $\|\eta(t,u)\| \ge \|u\| - |t|$. From this, together with $\eta(t,u) \in N^\perp$, it follows that the denominator of the vector field is not zero for $|t| < \|u\|$.

CLAIM. ξ *is well defined on* $B_\delta^N \times B_\delta^\perp$, where $B_\delta^N = B_\delta \cap N$, and $B_\delta^\perp = B_\delta \cap N^\perp$ for some $\delta > 0$. In fact, the following inequalities hold:

(a) $\forall\varepsilon > 0,\ \exists\delta_1 = \delta(\varepsilon) > 0$ such that

$$|F(u,y) - F_2(u)| = |F(u,y) - F(\theta_1,y) - (d_uF(\theta_1,y),u) - F_2(u)|$$
$$= \left|\int_0^1 (1-t)((d_u^2F(tu,y) - d_u^2F(\theta_1,\theta_2))u,u)\,dt\right|$$
$$< \varepsilon\|u\|^2,$$

where $(u,y) \in B^N_{\delta_1} \times B^{\perp}_{\delta_1}$.

(b) $$|F_2(\eta(t,u)) - F_2(u)| = \left|\int_0^t \frac{d}{ds} F_2(\eta(s,u))\,ds\right|$$
$$= \left|\int_0^t (dF_2(\eta),\dot{\eta})\,ds\right|$$
$$= \int_0^{|t|} \|A\eta(s,u)\|\,ds$$
$$\geq C\int_0^{|t|} \|\eta(s,u)\|\,ds$$
$$\geq C\left(\|u\|\,|t| - \frac{t^2}{2}\right),$$

where $C > 0$ is a constant determined by the spectrum of A. We conclude that

(c) $F_2(\eta(t,u))$, as a function of t, is strictly decreasing on $(-\|u\|,\|u\|)$.

(d) $F_2(\eta(-t,u)) > F(u,y) > F_2(\eta(t,u))$

holds for

$$\left(1 - \sqrt{1 - \frac{2\varepsilon}{C}}\right)\|u\| \leq t \leq \|u\|. \tag{2.4}$$

Therefore, there exists a unique $\bar{t}(u,y)$ with

$$|\bar{t}(u,y)| \leq \left(1 - \sqrt{1 - \frac{2\varepsilon}{C}}\right)\|u\| \tag{2.5}$$

such that

$$F_2(\eta(\bar{t}(u,y),u)) = F(u,y). \tag{2.6}$$

Thus the function ξ is of the form:

$$\xi(u,y) = \begin{cases} \theta_1 & u = \theta_1 \\ \eta(\bar{t}(u,y),u) & u \neq \theta_1 . \end{cases}$$

3° Define a map $\Psi:(u,y) \mapsto (\xi(u,y),y)$. We shall verify that Ψ is a local homeomorphism. That $\bar{t}(u,y)$ is continuous, easily follows from the implicit function theorem for $u = u_0 \neq \theta$, provided

$$(2.7) \qquad \frac{\partial}{\partial t} F_2 \circ \eta(\bar{t},u_0) = -\|A\eta(\bar{t},u_0)\| \neq 0;$$

and for $u = \theta$, provided by (2.5).

We have used the path $\eta(t,u)$ to carry a point (u,y) to the point $(\xi(u,y),y)$; the same path can be used for the opposite purpose, i.e., to define the inverse map $\Phi = \Psi^{-1}$. The same reason is provided to verify the continuity of Φ. Therefore Φ is a homeomorphism.

The equality (2.1) follows directly from (2.6). □

In case $N = \{\theta\}$, the Morse lemma is a consequence of this theorem except the conclusion that Φ is a diffeomorphism.

PROOF of the Morse lemma. We have proved that Φ is a homeomorphism. Now we shall prove it is a diffeomorphism. That $\bar{t}(u)$, and then $\xi(u)$, is continuously differentiable for $u \in B_\delta \backslash \{\theta\}$, follows from the implicit function theorem and (2.7). It is also easily obtained that $d\xi(\theta) = \mathrm{id}$, by using

$$(2.8) \qquad \|\eta(\bar{t}(u),u) - u\| \le |\bar{t}(u)| \le \left(1 - \sqrt{1 - \frac{2\varepsilon}{C}}\right)\|u\| = o(\|u\|)$$

as $u \to \theta$. Hence, $\|d\xi(u) - \mathrm{id}\| = o(1)$ remains to be proved. Since

$$\eta(t,u) = u - \int_0^t \frac{A\eta(s,u)}{\|A\eta(s,u)\|} ds \quad \text{for} \quad |t| < \|u\|$$

we write $\eta(t) = \eta(t,u)$, $\eta_u(t) = d_u\eta(t,u)$ for simplicity,

$$(2.9) \qquad \eta_u(t) = \mathrm{id} - \int_0^t \left[\frac{A\eta_u(s)}{\|A\eta(s)\|} - \frac{A\eta(s) \otimes ((A\eta_u(s))^* A\eta(s))}{\|A\eta(s)\|^3}\right] ds$$

that is

$$\|\eta_u(t)\| \le 1 + C_1 \int_0^1 \frac{1}{\|\eta(s)\|} \|\eta_u(s)\| \, ds,$$

where $C_1 > 0$ is a constant depending on $\sigma(A)$. But

$$\sqrt{1 - \frac{2\varepsilon}{C}}\,\|u\| \le \|\eta(s)\| \le 2\|u\| \quad \text{as} \quad |s| \le |\bar{t}(u)|. \tag{2.10}$$

Applying the Gronwall inequality,

$$\|\eta_u(t)\| \le 1 + e^{\frac{2C_1}{\|u\|} t} \le C_2 \qquad \text{for} \quad 0 < |t| \le \bar{t}(u).$$

Thus

$$\|\eta_u(\bar{t}(u), u) - id\| \le C_3 \frac{|\bar{t}(u)|}{\|u\|} = o(1) \quad \text{as} \quad \|u\| \to 0.$$

Since

$$\|d\xi(u) - id\| = \|\eta_u(\bar{t}(u), u) - id + d_u\bar{t}(u) \cdot \eta_t'(\bar{t}(u), u)\|$$

and by (2.6) and (2.10), we have

$$\|d_u\bar{t}(u)\| = \frac{\|d_uF(u) - \eta_u(\bar{t}(u), u)A\eta(\bar{t}(u), u)\|}{\|A\eta(\bar{t}(u), u)\|}$$

$$= \frac{\|Au + o(\|u\|) - (id + o(1))A(u + o(\|u\|))\|}{\|A(u + o(\|u\|))\|} = o(1) \quad \text{as} \quad u \to \theta.$$

This proves that $\|d\xi(u) - id\| \to 0$ as $\|u\| \to 0$. The implicit function theorem is employed, ξ is a local diffeomorphism. □

REMARK 2.1. There is no difficulty in extending the Morse lemma as well as Theorem 2.1 to the case M = Hilbert Riemannian manifold. The theorem was first obtained by Gromoll and Meyer [39] under the condition $f \in C^3$. However, the C^2-Morse lemma was proved by N.H. Kuiper in [48] from which the idea of the above proof was taken. The C^3-Morse lemma was given by Palais [61], Nirenberg [59] and Schwartz [75].

Let M be a C^2 Finsler manifold, and let $f \in C^1(M, \mathbb{R}^1)$ be a function satisfying the P.S. condition. Suppose that $V: M\backslash K \to T(M)$ is a p.g.v.f. of f.

DEFINITION 2.1. Let p be an isolated critical point of f, $c = f(p)$. A pair of topological spaces (W, W_-) is called a *Gromoll-Meyer pair w.r.to* V, if

(1) W is a closed neighbourhood of p possessing the mean value property, i.e., $\forall t_1 < t_2$, $\eta(t_i) \in W$, $i = 1,2$, implies $\eta(t) \in W$ for all $t \in [t_1, t_2]$, where $\eta(t)$ is the decreasing flow w.r.to V. And, there exists $\varepsilon > 0$ such that $W \cap f_{c-\varepsilon} = f^{-1}[c-\varepsilon, c) \cap K = \emptyset$, $W \cap K = \{p\}$.

(2) $W_- := \{x \in W \mid \eta(t,x) \notin W \quad \forall t > 0\}$ is closed in W.

(3) W_- is a piecewise submanifold, and the flow η is transversal to W_-, $\eta \pitchfork W_-$.

At this moment, the existence of a Gromoll-Meyer pair is assumed. The following theorem claims the motivation of the definition.

THEOREM 2.2. *Let* (W, W_-) *be a Gromoll-Meyer pair w.r.to a p.g.v.f.* V *of an isolated critical point* p *of the function* $f \in C^{2-0}(M, \mathbb{R}^1)$ *satisfying the P.S. condition. Then we have*

$$H_*(W, W_-; G) \simeq C_*(f, p).$$

PROOF. We introduce two sets $U_+ = \bigcup_{0<t<\infty} \eta(t, W)$, and $\tilde{U}_+ = \bigcup_{0<t<\infty} \eta(t, W_-)$, and define a continuous function $\gamma(x)$ on $\tilde{U}_+$ such that $\eta(\gamma(x), x) \in f_{c-\varepsilon} \cap \tilde{U}_+$, but $\eta(t,x) \notin f_{c-\varepsilon} \cap \tilde{U}_+$ if $t < \gamma(x)$. Thus

$$\sigma(t,x) = \eta(t\gamma(x), x) \qquad t \in [0,1], \quad x \in \tilde{U}_+,$$

defines a strong deformation retract $\tilde{U}_+ \to f_{c-\varepsilon} \cap \tilde{U}_+ = f_{c-\varepsilon} \cap U_+$, which reduces to

$$H_*(f_{c+\varepsilon} \cap U_+, f_{c-\varepsilon} \cap U_+; G) \simeq H_*(U_+, U_+; G).$$

Noticing that $\eta \pitchfork W_-$, $\forall y \in U_+$, there are unique $x \in W_-$, $t \geq 0$, such that $y = \eta(t,x)$ and $t = t(y)$, $x = x(y)$ continuously depend on y. The mean value property implies that $U_+ = W \cup \bigcup_{0<t<\infty} \eta(t, W_-)$.

Let $\tilde{U}_\delta = \bigcup_{\delta<t<+\infty} \eta(t,W_-)$, for $\delta > 0$, i.e., $\tilde{U}_\delta = \{y \in U_+ \mid \hat{t}(y) > \delta\}$, where we define $\hat{t}(y) = 0 \ \forall y \in W$, and $\hat{t}(y) = t(y) \ \forall y \in \tilde{U}_+$. Hence $\tilde{U}_\delta$ is open,

$$\overline{\tilde{U}_\delta} = \bigcup_{\delta\le t<+\infty} \eta(t,W_-) \subset \bigcup_{0<t<+\infty} \eta(t,W_-) = \mathrm{int}(\tilde{U}_+).$$

By the excision property, we have

$$H_*(U_+,\tilde{U}_+;G) \cong H_*(U_+\backslash\tilde{U}_\delta,\tilde{U}_+\backslash\tilde{U}_\delta;G).$$

Again, there is a strong deformation retract, defined by the reversed flow $\eta(-t):((U_+\backslash\tilde{U}_\delta),(\tilde{U}_+\backslash\tilde{U}_\delta)) \to (W,W_-)$, provided $\eta \pitchfork W_-$, we obtain

$$H_*(U_+\backslash\tilde{U}_\delta,\tilde{U}_+\backslash\tilde{U}_\delta;G) \cong H_*(W,W_-;G).$$

On the other hand, by the deformation property,

$$H_*(f_{c+\varepsilon} \cap U_+, f_{c-\varepsilon} \cap U_+;G) \cong H_*(f_c \cap U_+,(f_c\backslash\{p\}) \cap U_+;G) \cong C_*(f,p).$$

Combining all these isomorphisms we obtain $H_*(W,W_-;G) \cong C_*(f,p)$. □

We now turn to construct a Gromoll-Meyer pair for an isolated critical point. For simplicity, we may assume that M is a Hilbert space, and $p = \theta$, $f(\theta) = 0$. Choose $\varepsilon > 0$, $\delta > 0$ such that 0 is the unique critical value in $[-\varepsilon,\varepsilon]$, and θ is the unique critical point in the ball B_δ centered at θ. Define a function

$$g(x) = \lambda f(x) + \|x\|^2,$$

a neighbourhood

(2.11) $$W = f^{-1}[-\gamma,\gamma] \cap g_\mu,$$

and a subset of W:

(2.12) $$W_- = f^{-1}(-\gamma) \cap W,$$

where λ, μ and γ are positive numbers to be determined by the following conditions:

$$B_{\delta/2} \cap f^{-1}[-\gamma,\gamma] \subset W \subset B_\delta \cap f^{-1}[-\varepsilon,\varepsilon] \tag{2.13}$$

$$f^{-1}[-\gamma,\gamma] \cap g^{-1}(\mu) \subset B_\delta \backslash B_{\delta/2} \tag{2.14}$$

$$(dg(x),df(x)) > 0 \qquad \forall x \in B_\delta \backslash \mathring{B}_{\delta/2}. \tag{2.15}$$

Due to the P.S. condition, there exists $\beta = \inf_{x \in B_\delta \backslash B_{\delta/2}} \|df(x)\| > 0$. λ, μ and γ are then determined consecutively: $\lambda > \frac{2\delta}{\beta}$, $0 < \gamma < \min\{\varepsilon, \frac{3\delta^2}{8\lambda}\}$, and $\frac{\delta^2}{4} + \lambda\gamma \le \mu \le \delta^2 - \lambda\gamma$.

THEOREM 2.3. *The pair* (2.11), (2.12) *is a Gromoll-Meyer pair w.r.to the negative gradient vector field* $-df(x)$.

PROOF. We only want to verify:

(1) the mean value property,

(2) $W_- = \{x \in W \mid \eta(t,x) \notin W, \ \forall t > 0\}$, for the negative gradient flow η.

CLAIM (1). For simplicity, we may assume $\eta(0), \eta(t) \in W$. We wish to prove $\eta(s) \in W \quad \forall s \in [0,t]$. Define $T = \sup\{s \in [0,t] \mid \eta(s') \in W \ \ \forall s' \le s\}$. If $T < t$, then $\eta(T) \notin \mathring{B}_{\delta/2}$. But

$$\gamma \ge f \circ \eta(0) \ge f \circ \eta(T) > f \circ \eta(t) \ge -\gamma$$

$$(g \circ \eta)'(T) = -(dg,df)/\|df\|^2\Big|_{\eta(T)} < 0$$

provided by (2.13) and (2.15). This contradicts the fact that T is a supremum. Therefore $T = t$.

CLAIM (2). We write $W^- = \{x \in W \mid \eta(t,x) \notin W, \ \forall t > 0\}$. Obviously we see $W_- \subset W^-$. Now we prove $W^- \subset W_-$. By definition, $W^- \subset \partial W$ and $\partial W = W_- \cup (f^{-1}(\gamma) \cap \mathring{g}_\mu) \cup (g^{-1}(\mu) \cap (W \backslash W_-))$. If $x \in f^{-1}(\gamma) \cap \mathring{g}_\mu$, then $x \notin W^-$. For $x \in g^{-1}(\mu) \cap (W \backslash W_-)$, according to (2.14) and (2.15), we have $(g \circ \eta)'(\theta,x) < 0$ and $f(x) > -\gamma$. These imply that $\exists \tau > 0$ such that $g \circ \eta(\tau,x) \le \mu$ and $|(f \circ \eta)(\tau,x)| \le \gamma$, i.e., $x \notin W^-$. We have proved $W_- = W^-$. □

We intend to compute the critical groups of an isolated critical point p, which may be degenerate; the splitting theorem is employed. Let Φ be a local parametrization of M defined in some open neighbourhood U if θ in $T_p(M) \cong H$ with $\Phi(\theta) = p$, such that $f \circ \Phi(z,y) = \frac{1}{2}(Az,z) + f_0(y)$, where $A = d^2f(p)$, 0 is either an isolated point of the spectrum $\sigma(A)$ or does not belong to $\sigma(A)$, and f_0 is a function defined on N, the null space of A. We call $\mathcal{N} = \Phi(U \cap N)$ the characteristic submanifold of M for f at p w.r. to the parametrization Φ. The following theorem sets up the relationship between the critical groups of f and those of $\tilde{f} := f|_{\mathcal{N}}$.

THEOREM 2.4 (Shifting theorem). *Assume that the Morse index of* f *at* p *is* j, *then we have*

$$C_q(f,p) \cong C_{q-j}(\tilde{f},p), \quad q = 0,1,\ldots \qquad (C_q \cong 0 \ \textit{for} \ q < 0).$$

First we need:

LEMMA 2.1. *Suppose that* $H = H_1 \oplus H_2$, $g_i \in C^2(H_i,\mathbb{R}^1)$, θ_i *is an isolated critical point of* g_i, $i = 1,2$, *where* H_1,H_2 *are Hilbert spaces. Assume that* (W_i,W_{i-}) *is a Gromoll-Meyer pair of* θ_i *w.r.to the gradient vector field of* g_i, $i = 1,2$; *then* $(W_1 \times W_2,(W_{1-} \times W_2) \cup (W_1 \times W_{2-}))$ *is a Gromoll-Meyer pair of the function* $f = g_1 + g_2$ *at* $\theta = \theta_1 + \theta_2$ *w.r.to the gradient vector field* df, *if* $f(\theta)$ *is an isolated critical value of* f.

This is easy to check. We omit the proof.

THEOREM 2.5. *Under the assumptions of Lemma* 2.1, *we have*

$$C_*(f,\theta) = C_*(g_1,\theta_1) \otimes C_*(g_2,\theta_2).$$

PROOF. This is a combination of the Theorem 2.2, Lemma 2.1 and the Künneth formula.

PROOF of Theorem 2.4. This is a combination of Theorems 2.1, 2.5 and 1.1.

REMARK 2.2. Theorem 2.5 was conjectured by Gromoll and Meyer [39], and was solved by G. Tian [79]. In [32], Dancer independently proved the conjecture in the finite dimensional case.

The Gromoll-Meyer theory can be extended naturally to isolated critical sets.

Let S be an isolated critical set of f with $S = K \cap \tilde{\mathcal{O}} \cap f^{-1}(\alpha,\beta)$ (cf. Section 1). A pair of topological spaces (W,W_-) is called a Gromoll-Meyer pair with respect to a pseudo-gradient vector field V of f, if

(1) W is a closed neighbourhood of S possessing the mean value property with $W \cap f_\alpha = W_+ \cap f^{-1}(\alpha) \cap K = \emptyset$, and $W \cap K = S$, where $W_+ = \bigcup_{0 \le t < \infty} \eta(t,W)$ and η is the negative pseudo-gradient flow.

(2) $W_- := \{x \in W \mid \eta(t,x) \notin W \ \forall t > 0\}$ is closed in W.

(3) W_- is a piecewise submanifold, and the flow η is transversal to W_-, $\eta \pitchfork W_-$.

By the definition of isolated critical sets, the Gromoll-Meyer pair (W,W_-) for an isolated critical set S of a function $f \in C^1(M,\mathbb{R}^1)$ satisfying the P.S. condition with $S = K \cap \tilde{\mathcal{O}} \cap f^{-1}(\alpha,\beta)$, with respect to a pseudo-gradient vector field V is quite easy to construct:

$$W = \tilde{\mathcal{O}} \cap f^{-1}[\alpha',\beta'], \quad W_- = W \cap f^{-1}(\alpha'),$$

where $\alpha < \alpha' < \inf_{x \in S} f(x)$, $\beta > \beta' > \sup_{x \in S} f(x)$.

We have that W is bounded if $\mathcal{O}$ is bounded, and that:

THEOREM 2.6. *Suppose that* (W,W_-) *is a G.M. pair of* S, *then*

$$H_*(W,W_-;G) \simeq H_*(f_\beta \cap \tilde{\mathcal{O}}, f_\alpha \cap \tilde{\mathcal{O}};G).$$

The proof is the same as for Theorem 2.2.

REMARK 2.3. In the terminology of Conley's Morse theory [29], W is called the isolated block of S, and W_- is the exit set of W.

Section 3

MORSE THEORY ON FINSLER MANIFOLDS

The Morse theory has been established on Hilbert Riemannian manifolds in Section 1, where the Morse lemma holds, and the critical groups of a nondegenerate critical point are figured out. New difficulties in developing a Morse theory on a Banach space arise from two aspects. One is conceptual: the nondegeneracy in Hilbert spaces does not make sense in Banach spaces; because the Hessian $d^2f(x)$ of a function f is a bounded linear operator from the space to its dual, one cannot say that $d^2f(x)$ has a bounded inverse if the space is not isomorphic to its dual. The other difficulty is technical. The Morse lemma is no longer very useful, because it is not compatible with the Palais-Smale condition. For example, the quadratic functional $f(u) = \int_0^1 u^2(t)\,dt$ does not satisfy the P.S. condition on the space $L^p[0,1]$ with $p > 2$.

In this section, we present two kinds of Morse theories on Finsler manifolds. The first one is set up directly; passing over the Morse lemma, a handle body decomposition theorem is established. The second relies upon the Hilbert space theory, where the Banach space is presumed to be imbedded into a Hilbert space H, and the function f is presumed to be extendable C^2 on H. Under certain additional conditions, the Morse theory on H can be naturally reduced to that on the Banach space.

3.1. Morse theory without the Morse lemma

Let X be a Banach space with a differentiable norm $\|.\|$. In order to define the nondegeneracy of a critical point of a function f defined on a Finsler manifold M modeled on X, we need some preparations.

An *operator* $L \in \mathcal{B}(X)$, the Banach algebra of all bounded linear operators from X to X, is said to be *hyperbolic* if its spectrum $\sigma(L)$ is contained in two compact sets seperated by the imaginary axis. Denote $\sigma_{\pm} = \sigma(L) \cap H_{\pm}$, where $H_{\pm}$ is the right (left, resp.) open half plane, and $R_z = (zI-L)^{-1}$ ($R_z^* = (\bar{z}I-L^*)^{-1}$), the resolvent of L, defined on the resolvent set $\rho(L)$. According to Riesz-Nagy [67], the projection operator

$$P = -\frac{1}{2\pi i}\int_{\partial D} R_z \, dz$$

is well-defined for any domain D with a rectifiable boundary ∂D, which disjoints with the spectrum $\sigma(L)$. In particular, if we choose a compact domain $D_{\pm} \subset H_{\pm}$ containing $\sigma_{\pm}$, and let $P_{\pm}$ be the corresponding projection, then we have: $P_{\pm}^2 = P_{\pm}$, $P_+P_- = P_-P_+ = 0$ and $P_+ + P_- = I$. In this case, the Banach space X is decomposed into $X_+ \oplus X_-$, where $X_{\pm} = P_{\pm}X$. Thus $X^* = X_+^* \oplus X_-^*$. Let $L_{\pm} = L|_{X_{\pm}}$.

According to the functional calculus, the operator

$$S_{\pm} = \frac{1}{2\pi i}\int_{\partial D_{\pm}} \sqrt{\pm z}\, R_z \, dz$$

is well-defined, and commutes with $P_{\pm}$.

Let $b: X \times X \to \mathbb{R}^1$ be a bilinear continuous form. We say that b is commutative with L if

$$b(Lu,v) = b(u,Lv). \tag{3.1}$$

By definition, let $B: X \to X^*$ be the corresponding linear operator of b. Then (3.1) means $BL = L^*B$. If b commutes with L, then we have

$$\begin{cases} BR_z = R_z^*B \\ BP_\pm = P_\pm^*B \\ BS_\pm = S_\pm^*B \end{cases}$$

because R_z is expressible as a power series in L.

LEMMA 3.1. *Let* b *be a bilinear continuous form defined on* $X \times X$. *Suppose that* b *commutes with a hyperbolic operator* L *satisfying* $b(Lu,u) > 0$ $\forall u \neq \theta$. *Then* b *is positive (negative) on* X_+ *(*X_-*, resp.); and*

$$b(x,x) = b(x_+,x_+) + b(x_-,x_-) \qquad \forall x \in X, \quad x_\pm = P_\pm x.$$

PROOF. Straightforward by calculation.

LEMMA 3.2 (cf. Uhlenbeck [82]). *Let* L *be a hyperbolic operator on* X, *with* $X_\pm$ *the positive (negative) invariant subspaces. Then there exists an equivalent norm* $|\!|\!|.|\!|\!|$ *with the same smoothness as the norm* $\|.\|$, *and a constant* $\rho > 0$ *such that*

$$|\!|\!|x_+ + x_-|\!|\!| = |\!|\!|x_+|\!|\!| + |\!|\!|x_-|\!|\!| \qquad \forall (x_+,x_-) \in X_+ \oplus X_-,$$
$$|\!|\!|e^{Lt}x_\pm|\!|\!| \geq (1 + \rho|t|)|\!|\!|x_\pm|\!|\!| \quad \text{for} \quad \pm t > 0.$$

PROOF. We only want to define the equivalent norm on X_+ and on X_- respectively. Since the exponential operator e^L is invertible on X_+,

$$\|e^{-Lt}x_+\| = \left\|\int_{\partial D_+} e^{-zt}(zI-L)^{-1}x_+ \, dz\right\|$$
$$\leq e^{-\delta_o t} \frac{|\partial D_+|}{\text{dist}(\partial D_+,\sigma_+)} \|x_+\| = Me^{-\delta_o t}\|x_+\|,$$

where δ_o is the distance between ∂D_+ and the imaginary axis, M is a constant. Letting $\rho_o = \frac{1}{M}$ and $k_o = e^{\delta_o}$, we obtain

$$\|e^{Lt}x_+\| \geq \rho_o k_o^t \|x_+\| \qquad \forall t > 0. \tag{3.2}$$

Define

$$\|x_+\| = \int_0^N \|e^{Ls}x_+\|\, ds,$$

where N is so chosen that $\rho_o k_o^N = \lambda > 1$. This is an equivalent norm with the same smoothness. However,

$$\begin{aligned}
\|e^{Lt}x_+\| &= \int_0^N \|e^{L(s+t)}x_+\|\, ds \\
&= \int_t^{N+t} \|e^{Ls}x_+\|\, ds = \int_0^N + \int_N^{N+t} - \int_0^t \\
&= \|x_+\| + \int_0^t [\|e^{LN}(e^{Ls}x_+)\| - \|e^{Ls}x_+\|]\, ds \\
&\geq \|x_+\| + (\lambda-1)\int_0^t \|e^{Ls}x_+\|\, ds.
\end{aligned}$$

Due to (3.2)

$$\inf_{s\geq 0} \|e^{Ls}x_+\| \geq \rho_o\|x_+\| \geq \rho_1\|x_+\| \quad \text{for some } \rho_1 > 0$$

so that we have

$$\|e^{Lt}x_+\| \geq (1 + (\lambda-1)\rho_1 t)\|x_+\|.$$

Similarly, we obtain an inequality for x_- and define $\|x_-\|$. The proof is complete. □

LEMMA 3.3. *For all* $\mu > 0$ *we have*

(3.3) $\langle d\|x_+\| - \mu d\|x_-\|, Lx\rangle \geq \rho \cdot \min\{1,\mu\}\|x\| \qquad \forall (x_+,x_-) \in (X_+\backslash\{\theta\}) \oplus (X_-\backslash\{\theta\})$.

PROOF. Since $d\|x_\pm\| \in X_\pm^*$, we have

$$\langle d\|x_+\|, Lx_+\rangle = \lim_{t\to+0} \frac{1}{t}(\|e^{Lt}x_+\| - \|x_+\|) \geq \rho\|x_+\| \qquad \forall x_+ \neq \theta$$

$$-\langle d\|x_-\|, Lx_-\rangle = \lim_{t\to+0} \frac{1}{t}(\|e^{-Lt}x_-\| - \|x_-\|) \geq \rho\|x_-\| \qquad \forall x_- \neq \theta$$

and

$$\langle d\|x_\pm\|, Lx_\mp\rangle = 0.$$

Thus

$$\langle d\|x_+\| - \mu d\|x_-\|, Lx\rangle \geq \rho(\|x_+\| + \mu\|x_-\|) \geq \rho \min\{1,\mu\}\|x\|. \square$$

Next, we turn to define the nondegenerary.

DEFINITION 3.1. Let M be a C^2 Finsler manifold, $f \in C^2(M,\mathbb{R}^1)$, let p_0 be an isolated critical point. We say that p_0 is *nondegenerate*, if there is a neighbourhood U of p_0 on which $T(M)$ is trivialized to be $U \times X$, such that there exists a hyperbolic operator $L \in \mathcal{B}(X)$ satisfying

(1) $d^2f(p_0)(Lx,y) = d^2f(p_0)(x,Ly) \quad \forall x,y \in X$

(2) $d^2f(p_0)(Lx,x) > 0 \quad \forall x \in X\backslash\{\theta\}$

(3) $\langle df(p),Lx\rangle > 0 \quad \forall p \in U, \quad x \in X\backslash\{\theta\}$

The dimension of the negative invariant subspace X_- of L is called the *index* of p_0.

By Lemma 3.1 the index does not depend upon the special choice of L.

The main result in this section is the following handle-body decomposition theorem.

THEOREM 3.1. *Suppose that* $f \in C^2(M,\mathbb{R}^1)$ *satisfies the P.S. condition, and that* p_0 *is a nondegenerate critical point with finite index* j. *Then there exists a local chart* (U,ψ) *of* p_0, *a homeomorphism* h *from the* j-*disk* D^j *into* $X_-\times\{\theta_+\}$, *the negative invariant subspace of* L, *a neighbourhood* $V \subset U$, *and an* $\varepsilon > 0$ *such that*

$$f_c \cap V \cong (f_{c-\varepsilon} \cap V) \cup \psi^{-1} \circ h(D^j)$$

with

$$f_{c-\varepsilon} \cap \psi^{-1} \circ h(D^j) = f^{-1}(c-\varepsilon) \cap \psi^{-1} \circ h(D^j) = \psi^{-1} \circ h(\partial D^j),$$

where $c = f(p_0)$.

On a local chart, the function f is regarded as a function defined on the Banach space X, and then the critical point p_0 moves to the origin θ.

LEMMA 3.4. *Suppose that* f *is a* C^2 *function on a Banach space* X.

Suppose that θ *is a nondegenerate finite index critical point of* f *with a hyperbolic operator* L. *Then there is a cone neighbourhood*

$$C_{\mu,\delta} = \{(x_+,x_-) \in X_+ \oplus X_- \mid \|x_+\| \le \mu\|x_-\|,\ \|x_-\| \le \delta\},$$

where $\mu,\delta > 0$, *such that*

(1) $\langle df(x),x_-\rangle < -\frac{\nu}{2}\|x_-\|^2$ *for some* $\nu > 0$,

(2) $\langle df(x),Lx\rangle > \frac{\nu}{2}\|x_-\|^2$,

(3) $f(x) < f(\theta) - \frac{\nu}{4}\|x_-\|^2$.

for all $x \in C_{\mu,\delta}\setminus\{\theta\}$.

PROOF. Since $\dim X_- < +\infty$ and $d^2f(\theta)$ is negative on X_-, we have $\nu > 0$ such that

$$d^2f(\theta)(x_-,x_-) \le -\nu\|x_-\|^2,$$
$$d^2f(\theta)(Lx_-,x_-) \ge \nu\|x_-\|^2.$$

But $d^2f(\theta)$ is bounded, hence $\exists M > 1$ such that

$$d^2f(\theta)(x_+,x_+) \le M\|x_+\|^2.$$

Without loss of generality, we may assume $\|L\| \le 1$.

(1) For given positive $\varepsilon < \dfrac{\nu}{8M(1+\mu)^2}$, $\exists\delta > 0$ such that

$$\|df(x) - d^2f(\theta)x\| < \varepsilon\|x\| \quad \text{for} \quad 0 < \|x\| < \delta.$$

This implies

$$\langle df(x),x_-\rangle - d^2f(\theta)(x_-,x_-) \le \varepsilon\|x\|\|x_-\| \le (1+\mu)\varepsilon\|x_-\|^2, \qquad \forall x \in C_{\mu,\delta}.$$

Thus

$$\langle df(x),x_-\rangle < -\frac{\nu}{2}\|x_-\|^2 \qquad \forall x \in C_{\mu,\delta}\setminus\{\theta\}.$$

(2) $\langle df(x),Lx\rangle - d^2f(\theta)(Lx,x) \ge -\varepsilon\|x\|^2$

$$\ge -\varepsilon(1+\mu)^2\|x_-\|^2 \qquad \forall x \in C_{\mu,\delta}.$$

Thus

$$<df(x),Lx> > \frac{\nu}{2}\|x_-\|^2 \qquad \forall x \in C_{\mu,\delta}\backslash\{\theta\}.$$

(3) From

$$|f(x) - f(\theta) - \tfrac{1}{2}d^2f(\theta)(x,x)| \le \varepsilon\|x\|^2 \quad \text{for } \|x\| < \delta,$$

it follows that

$$f(x) \le f(\theta) - \frac{\nu}{4}\|x_-\|^2, \qquad \forall x \in C_{\mu,\delta}\backslash\{\theta\}$$

with $\mu < \min\{1,\sqrt{\frac{\nu}{4M}}\}$. □

In the sequel, let us denote by $\sigma(t,\cdot)$ the pseudo-gradient flow

$$\dot{\sigma}(t,x) = -\frac{\Xi(\sigma(t,x))}{<df,\Xi> \circ \sigma(t,x)}, \qquad \sigma(0,x) = x$$

and $\tilde{W} = \bigcup_{t\ge 0} \sigma(t,W)$ for each subset W. Assume $f(p_o) = 0$.

LEMMA 3.5. *For a given* $r > 0$ *suppose that* f *has only one critical point* p_o *in the ball* $B(p_o,2r)$. *Then there is an* $\varepsilon > 0$ *such that*

$$(\widetilde{f_o \cap B(p_o,r)})\backslash f_{-\varepsilon} \subset B(p_o,2r).$$

PROOF. If not, $\exists x_o \in B(p_o,r) \cap f_o$, $t_o > 0$ such that $\sigma(t_o,x_o) \notin B(p_o,2r)$, and $f \circ \sigma(t) > -\varepsilon$. Since there is a positive number $d > 0$ such that

$$\|df(x)\| \ge d \qquad \forall x \in B(p_o,2r)\backslash B(p_o,r)$$

provided by the P.S. condition, we choose $\varepsilon < \frac{dr}{2}$. Then we have $t_1,t_2 \in [0,t_o]$ such that $\sigma(t_1,x_o) \in \partial B(p_o,r)$, $\sigma(t_2,x_o) \in \partial B(p_o,2r)$. It follows that

$$t_2 - t_1 = -\int_{t_1}^{t_2} <df,\dot{\sigma}>\, ds = -\int_{t_1}^{t_2} \frac{d(f\circ\sigma)(s)}{ds}\, ds$$

$$= f \circ \sigma(t_1,x_o) - f \circ \sigma(t_2,x_o) < \varepsilon$$

and

$$\gamma \le \mathrm{dist}(\sigma(t_2,x_o),\sigma(t_1,x_o)) \le \int_{t_1}^{t_2} \|\dot{\sigma}(s)\|\, ds \le \frac{2}{d}(t_2-t_1) \le \frac{2\varepsilon}{d} < \gamma.$$

This is a contradiction. □

In proving Theorem 3.1, we choose $V = B(p_o,r)$, for suitable $r > 0$ such that $B(p_o,2r) \subset U$.

PROOF of Theorem 3.1. We may assume that f is defined on X. The proof is divided into two steps:

Firstly, there exist μ, ν and $\delta > 0$, as shown in Lemma 3.4, such that $(f_{-\varepsilon} \cap V) \cup C_{\mu,\delta}$ is a deformation retract of $f_o \cap V$, for $0 < \varepsilon < \frac{\nu}{4}\delta^2$. Secondly, $(f_{-\varepsilon} \cap V) \cup (D_\delta^- \times \{\theta_+\})$ is a deformation retract of $(f_{-\varepsilon} \cap V) \cup C_{\mu,\delta}$ where D_δ^- is the disk of radius δ in X_-.

1° Choose $\delta > 0$ so small that $D_\delta^- \times D_{\mu\delta}^+ \subset V$ and $C_{\mu,\delta} \cap (\partial D_\delta^- \times X_+) \subset \mathring{f}_{-\varepsilon}$. We have a smaller δ_1, $0 < \delta_1 < \delta$, such that $C_{\mu,\delta} \cap (\partial D_{\delta_1}^- \times X_+) \subset \mathring{f}_{-\varepsilon}$. We can define a C^{1-0} function

$$\rho_o(x) = \begin{cases} 1, & \|x_-\| \geq \delta \text{ or } \|x_+\| \geq \mu\delta \\ 0, & \|x_-\| \leq \delta_1 \text{ and } \|x_+\| \leq \mu\delta_1 \end{cases}$$

which is linear in $\|x_+\|$ and $\|x_-\|$, in $(D_\delta^- \times D_{\mu\delta}^+)\backslash(D_{\delta_1}^- \times D_{\mu\delta_1}^+)$. According to the p.g.v.f. Ξ of f, we define a new vector field

$$X(x) = (1 - \rho_o(x))Lx + \rho_o(x)\Xi(x).$$

For $x \in (f_o \cap V)\backslash\{\theta\}$ the flow

$$\dot{\sigma}(t) = -X(\sigma), \qquad \sigma(0) = x$$

is well defined and $\sigma(t) \in (f_o \cap V)$. We shall prove that σ applies for the deformation. Let $[0,\beta)$ be the maximal interval for the existence of the flow. Then after a finite time, say $\gamma_x < +\infty$, we have $\sigma(\gamma_x,x) \in (f_{-\varepsilon} \cap V) \cup C_{\mu,\delta}$ $\forall x \in (f_o \cap V)\backslash\{\theta\}$. This is proved in the following

<u>Case I</u>. $\inf \operatorname{dist}(D_{\delta_1}^- \times D_{\mu\delta_1}^+, \sigma(t,x)) > 0$.

$\exists\varepsilon_1 > 0$ such that $\rho_o(\sigma(t,x)) \geq \varepsilon_1$. We have

$$d(f\circ\sigma)(t) = -\langle df,\chi\rangle \circ \sigma(t) \leq -\varepsilon_1\langle df,\Xi\rangle \circ \sigma(t) \leq -\varepsilon_1 d^2$$

provided by (3) in Definition 3.1. Thus,

$$f(x) - f\circ\sigma(t) \geq \varepsilon_1 t d^2.$$

Then either $\lim_{t\to\beta} f\circ\sigma(t) = -\infty$, or β is finite and $\int_0^\beta \|df(\sigma(t))\|^2\,dt < \infty$. In the first case, $\exists T \in [0,\beta)$ such that $\sigma(T,x) \in f_{-\varepsilon} \cap V$. In the second case,

$$\begin{aligned}\int_0^\beta \|\dot\sigma(s)\|\,ds &= \int_0^\beta \|\chi(\sigma(s))\|\,ds\\ &\leq \int_0^\beta \|\Xi(\sigma(s))\|\,ds + \int_{\substack{\sigma(s)\in D_\delta^-\times D_{\mu\delta}^+\\ 0\leq s\leq\beta}} \|L\sigma(s)\|\,ds\\ &\leq 2\int_0^\beta \|df(\sigma(s))\|\,ds + \beta M\\ &\leq 2\{\beta\int_0^\beta \|df(\sigma(s))\|^2\,ds\}^{\frac{1}{2}} + \beta M < +\infty\end{aligned}$$

where M is a constant depending on $\|L\|$ and δ. Then the limit $\lim_{t\to\beta}\sigma(t,x)$ exists in $f_o\backslash D_{\delta_1}^-\times D_{\mu\delta_1}^+$, which contradicts the maximality of β.

Case II. $\inf_{t\in[0,\beta)} \operatorname{dist}(D_{\delta_1}^- \times D_{\mu\delta_1}^+,\sigma(t,x)) = 0$. There are two possibilities: either $\lim_{t\to\beta}\operatorname{dist}(D_{\delta_1}^- \times D_{\mu\delta_1}^+,\sigma(t,x)) = 0$ or there exists $\delta_2 \in (0,\frac{1}{2}(\delta-\delta_1))$ and infinitely many disjoint intervals $[\alpha_j,\beta_j]$, $j = 1,2,\ldots$, such that

$$\sigma(t,x) \in (D_{\delta_1+2\delta_2}^- \times D_{\mu(\delta_1+2\delta_2)}^+)\backslash(D_{\delta_1+\delta_2}^- \times D_{\mu(\delta_1+\delta_2)}^+) \quad \forall t \in [\alpha_j,\beta_j],\ j = 1,2,\ldots$$

However, the latter case cannot happen, except when $\sigma(t,x)$ enters into $f_{-\varepsilon} \cap V$ after a finite time. In fact, there is a $d_1 > 0$ such that

$$\rho_o(\sigma(t,x)) \geq d_1 \quad \forall t \in [\alpha_j,\beta_j],\ j = 1,2,\ldots .$$

On the one hand,

$$f(x) - \lim_{t\to\beta} f \circ \sigma(t,x) \geq \sum_j \int_{\alpha_j}^{\beta_j} \langle df,\chi\rangle \circ \sigma(t)\, dt \geq d_2^2 d_1 \sum_j (\beta_j - \alpha_j),$$

where $d_2 = \inf\limits_{x\in(D_\delta^- \times D_{\mu\delta}^+)\setminus(D_{\delta_1}^- \times D_{\mu\delta_1}^+)} \|df(x)\|$, which implies $\sum_j (\beta_j - \alpha_j) < +\infty$. On the other hand,

$$\delta_2 \leq \|\sigma(\beta_j,x) - \sigma(\alpha_j,x)\| \leq \int_{\alpha_j}^{\beta_j} \left\|\frac{d\sigma}{d\tau}\right\| d\tau \leq M(\beta_j - \alpha_j).$$

This is a contradiction.

Now we turn to the former case: $\lim\limits_{t\to\beta} \operatorname{dist}(D_{\delta_1}^- \times D_{\mu\delta_1}^+, \sigma(t,x)) = 0$. According to Lemma 3.2, after a finite time, $\sigma(t,x)$ enters into $C_{\mu,\delta}$, if one can prove that there exists a finite $T \in [0,\beta)$ such that $\sigma(T,x) \in D_{\delta_1}^- \times D_{\mu\delta_1}^+$. Suppose there is no such T. Since for each $\delta_3 > \delta_1$, $\exists T_1 \in [0,\beta)$ such that

$$\sigma(t,x) \in D_{\delta_3}^- \times D_{\mu\delta_3}^+ \setminus D_{\delta_1}^- \times D_{\mu\delta_1}^- \qquad \forall t > T_1$$

and

$$\|\Xi(x)\| \leq 2\|df(x)\| \leq 2\|d^2 f(\theta)x\| + o(\|x\|),$$

it follows that

$$\|L\sigma - \Xi(\sigma)\| \leq M_1 \|\sigma(t)\|,$$

where M_1 is a constant. Observing the equation

$$\dot{\sigma}(t) = -L\sigma + \rho_o(\sigma)[L\sigma - \Xi(\sigma)],$$

we obtain

$$\sigma(t+\Delta t) = e^{-L\Delta t}\sigma(t) + \int_t^{t+\Delta t} e^{-L(t+\Delta t-\tau)} \rho_o(\sigma)[L\sigma - \Xi(\sigma)]\, dt.$$

Hence,

$$\begin{aligned} &\|P_-\sigma(t+\Delta t)\| - \|P_-\sigma(t)\| \\ &\quad \geq \|e^{-L\Delta t} P_-\sigma(t)\| - \|P_-\sigma(t)\| - \varepsilon_3 M_1 \int_t^{t+\Delta t} \|e^{-L(t+\Delta t-\tau)}\|\, d\tau\, \|\sigma(t)\|, \end{aligned}$$

where $\varepsilon_3 = \sup_{x \in D^-_{\delta_3} \times D^+_{\mu\delta_3}} \rho_o(x)$. By Lemma 3.2,

$$\frac{d}{dt} \| P_- \sigma(t) \| \geq \rho \| P_- \sigma(t) \| - \varepsilon_3 M_1 \| \sigma(t) \| .$$

Similarily, we have

$$\frac{d}{dt} \| P_+ \sigma(t) \| \leq -\rho \| P_+ \sigma(t) \| + \varepsilon_3 M_1 \| \sigma(t) \| .$$

Choosing $\delta_3 - \delta_1 > 0$ sufficiently small such that $\varepsilon_3 < \frac{\mu\rho}{2M_1}$, we obtain

$$\frac{d}{dt} (\| P_- \sigma(t) \| - \| P_+ \sigma(t) \|) \geq \alpha \| \sigma(t) \| \qquad \forall t > T_1$$

where $\alpha = \rho - 2M_1 \varepsilon_3 > 0$. Then we arrive at the inequality

$$(3.4) \quad (\| P_- \sigma(t) \| - \| P_+ \sigma(t) \|) - (\| P_- \sigma(T_1) \| - \| P_+ \sigma(T_1) \|) \geq \alpha \int_{T_1}^{t} \| \sigma(\tau) \| \, d\tau .$$

But (3.4) cannot hold, because the LHS of (3.4) is finite, and $\| \sigma(t) \| \geq \mu\delta_1$ for $t > T_1$; these imply $\beta < +\infty$. Then

$$\int_{T_1}^{\beta} \| \dot{\sigma}(\tau) \| \, d\tau = \int_{T_1}^{\beta} \| X(\sigma(\tau)) \| \, d\tau \leq M \int_{T_1}^{\beta} \| \sigma(\tau) \| \, d\tau < +\infty$$

which implies that $\lim_{t \to \beta} \sigma(t)$ exists in $(f_o \cap V) \backslash (D^-_{\delta_1} \times D^+_{\mu\delta_1})^o$. This is a contradiction.

If we denote by γ_x the first arriving time of $\sigma(t,x)$ in $(f_{-\varepsilon} \cap V) \subset C_{\mu,\delta}$, we shall prove that the function $x \mapsto \gamma_x$ is continuous in $f_o \cap V \backslash \{\theta\}$. In fact, $\langle df(x), X(x) \rangle > 0 \ \forall x \in f_o \backslash \{\theta\}$ implies $X \pitchfork f^{-1}(-\varepsilon)$; and (3.3) implies $X \pitchfork \partial C_{\mu,\delta} \backslash (f_{-\varepsilon} \cup \{\theta\})$. Inside $D^-_{\delta_1} \times D^+_{\mu\delta_1}$, $\dot{\sigma} = -L\sigma$, which implies $\lim_{x \to \theta} \sigma(t\gamma_x, x) = \theta$ uniformly in $t \in [0,1]$. Hence the function

$$\tau(t,x) = \begin{cases} \sigma(t\gamma_x, x), & \text{if } x \in (f_o \cap V) \backslash \{\theta\}, \\ \theta \quad , & \text{if } x = \theta \end{cases}$$

defines a deformation retract of $f_o \cap V$ into $(f_{-\varepsilon} \cap V) \cup C_{\mu,\delta}$.

2° The second deformation retract $(f_{-\varepsilon} \cap V) \cup C_{\mu,\delta} \to (f_{-\varepsilon} \cap V) \cap (D_\delta^- \times \{\theta_+\})$ is constructed in two steps.

Firstly, $\exists \delta_o, k_o > 0$ such that $(f_{-\varepsilon} \cap V) \cap C_{\mu,\delta} \subset T_{k_o,\delta_o} := \{(x_+, x_-) \in C_{\mu,\delta} \mid \|x_-\| \geq \frac{\|x_+\|}{k_o} + \delta_o\}$. In fact, $\exists M_1 > 0$ such that for $\delta > 0$ small, we have:

$$-\varepsilon > f(x) > d^2f(\theta)(x,x) - \|x_+\|^2 - \|x_-\|^2 \geq -M_1\|x_-\|^2 - \|x_+\|^2$$
$$\geq -(M+2\mu)\|x_-\|^2 + \|x_+\|^2.$$

One defines a deformation as follows:

$$\eta_1(t,x) = \begin{cases} x & , \quad x \in f_{-\varepsilon} \cap V \\ x_- + \left(t\,\dfrac{k_o(\|x_-\|-\delta_o)}{\|x_+\|} + (1-t)\right)x, & \quad x \in C_{\mu,\delta} \cap \{\delta_o \leq \|x_-\| \leq \dfrac{\|x_+\|}{k_o} + \delta_o\} \\ x_- + t(1-t)x_+ & , \quad x \in C_{\mu,\delta} \cap \{\|x_-\| \leq \delta_o\}. \end{cases}$$

It deforms $(f_{-\varepsilon} \cap V) \cup C_{\mu,\delta}$ into $(f_{-\varepsilon} \cap V) \cup T_{k_o,\delta_o} \cup \overline{D}_{\delta_o}^- \times \{\theta_+\}$.

Since for $(x_-,x_+) \in C_{\mu,\delta}\backslash\{\theta\}$, $\exists! r, 1 < \gamma = \gamma(x) < \frac{\delta}{\|x_-\|}$ such that $f(x_+ + \gamma x_-) = -\varepsilon$. The function γ is continuous, provided by (1) in Lemma 3.4. One constructs another deformation: $\forall (t,x) \in [0,1] \times (f_{-\varepsilon} \cap V) \cup C_{\mu,\delta}$,

$$\eta_2(t,x) = \begin{cases} x & , \quad x \in f_{-\varepsilon} \cap V \\ x_+ + x_- - t\left(\dfrac{b-c}{b-a}\right)(\|x_-\|-a)\,\dfrac{x_-}{\|x_-\|} & , \quad a \leq \|x_-\| \leq b \\ x_+ + (1-t)\left(1 - \dfrac{c}{\|x_-\|}\right)x_- + \dfrac{c}{\|x_-\|}\,x_- & , \quad \begin{cases} b \leq \|x_-\| \leq c & \text{if } a \leq b, \text{ or} \\ a \leq \|x_-\| \leq c & \text{if } a > b \end{cases} \end{cases}$$

where $a = \frac{1}{\mu}\|x_+\|$, $b = \frac{1}{k_o}\|x_+\| + \delta_o$, and $c = \gamma\|x_-\|$. It deforms $(f_{-\varepsilon} \cap V) \cup T_{k_o,\delta_o} \cup (\overline{D}_{\delta_o}^- \times \{\theta_+\})$ into $(f_{-\varepsilon} \cap V) \cup (D_\delta^- \times \{\theta_+\})$ in $(f_{-\varepsilon} \cap V) \cup C_{\mu,\delta}$. Since $\dim X_-$ is finite, and $f|_{X_-}$ has a local minimum at θ_-, we may choose $\varepsilon > 0$ so small that $X_- \cap f^{-1}[-\varepsilon,0)$ is radially homeomorphic to the disk D^- in X_-.

Composing η_1 with η_2, the theorem is proved. □

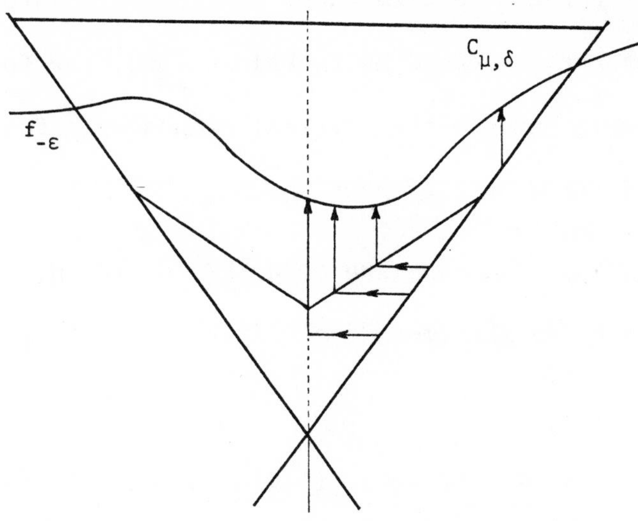

Directly, we arrive at

COROLLARY 3.1. *Assume that* p_o *is a nondegenerate critical point of* $f \in C^2(M,\mathbb{R}^1)$, *with index* $j < +\infty$*; then*

$$C_q(f,p_o) = \delta_{jq}G.$$

PROOF.
$$\begin{aligned} C_q(f,p_o) &= H_q(f_c \cap V, f_c \cap V\backslash\{p_o\};G) \\ &= H_q((f_{c-\varepsilon} \cap V) \cup \psi^{-1} \circ h(D^j), (f_{c-\varepsilon} \cap V) \cup \psi^{-1} \circ h(D^j\backslash\{\theta\});G) \\ &= H_q(D^j, D^j\backslash\{\theta\};G) \\ &= \delta_{jq}G. \ \square \end{aligned}$$

REMARK 3.1. Theorem 3.1 was given in Chang [22]. Originally, it was stated in K. Uhlenbeck [82] under a very weak definition of nondegeneracy, but the proof was incomplete.

REMARK 3.2. A different Morse theory on a Finsler manifold was given by Tromba [81]. The difference between his theory and the presentation above is two-fold: (1) He assumed a gradient-like vector field with some properties similar to the P.S. condition in replacing the P.S. condition on the function f. (2) A slightly different definition of nondegeneracy was given.

3.2. Hereditable theory

Let X be a Banach space imbedded continuously into a Hilbert space H as a dense linear subspace. According to Palais [62], the following theorem establishes the relationship between the critical groups of a function regarded as different functions defined on different spaces.

THEOREM 3.2 (Palais). *For any open set* O *in* H, *let* $\tilde{O} = O \cap X$, *and* $\tilde{i} = i|_{\tilde{O}}$ *where* $i: X \to H$ *is the imbedding. Then* $\tilde{i}: \tilde{O} \to O$ *is a homotopy equivalence.*

Thus, for a pair of open sets (A,B) in H, let $(A|_X, B|_X)$ be the restriction of (A,B) on X. We have

$$H_*(A,B;G) = H_*(A|_X, B|_X;G). \tag{3.5}$$

For a function $f \in C^1(H,\mathbb{R}^1)$ let $\tilde{f} = f|_X$, by definition; then $\tilde{f} \in C^1(X,\mathbb{R}^1)$. However, the P.S. condition of f cannot be transferred to $\tilde{f}$ in general. The following theorem deals with the inheritability of the deformation property.

Assume that $f \in C^{2-0}(H,\mathbb{R}^1)$ satisfies the P.S. condition, with gradient of the form

$$df(x) = x - \mathbb{K} \circ \mathbb{G}(x), \tag{3.6}$$

where $\mathbb{K}$ and $\mathbb{G}$ are operators satisfying the following assumptions:

(H_1) There are two chains of Banach spaces

$$X_N \hookrightarrow X_{N-1} \hookrightarrow \cdots \cdots \hookrightarrow X_1 \to X_0$$
$$Y_{N-1} \hookrightarrow Y_{N-2} \hookrightarrow \cdots \cdots \hookrightarrow Y_0$$

with $H \hookrightarrow X_0$, $X_N \hookrightarrow X$.

(H_2) Let $\|\cdot\|_i = \|\cdot\|_{X_i + X}$ (we may assume $\|\cdot\|_i \le \|\cdot\|_{i+1}$), $i = 0,1,\ldots,N$.

$\mathbb{G}: X_i \to X \to Y_i$ has the property that for every neighbourhood V in X there is an $M_i = M_i(V)$ such that

$$\|\mathbb{G}(x) - \mathbb{G}(y)\|_{Y_i} \leq M_i \|x - y\|_i \qquad \forall x,y \in V$$

and $\mathbb{K} \in L(Y_i, X_{i+1})$.

Furthermore, we assume

(H_3) The critical set K of f is in X.

An example in mind is the following BVP:

$$-\Delta u = g(x,u) \qquad \Omega \subset \mathbb{R}^n, \text{ bounded domain with smooth boundary.}$$
$$u\big|_{\partial\Omega} = 0$$

We assume that $g \in C^1(\bar{\Omega} \times \mathbb{R}^1, \mathbb{R}^1)$ satisfies the growth condition

$$|g_t'(x,t)| \leq C(1 + |t|^{\alpha-1})$$

where $\alpha < \frac{n+2}{n-2}$, and C is a constant. The functional

$$f(u) = \int_\Omega \left(\frac{|\nabla u|^2}{2} - G(x,u(x))\right) dx$$

where $G(x,\xi) = \int_0^\xi g(x,t)\, dt$, is defined on $H = H_0^1(\Omega)$, the Sobolev space, as well as on the space $X = C_0^1(\bar{\Omega})$, the continuously differentiable space with 0-boundary condition. In this example, $\mathbb{K} = (-\Delta)^{-1}$ and $\mathbb{G}$ is taken to be the Nemytski operator: $u \mapsto g(x,u(x))$, with

$$X_{i+1} = W^2_{p_{i+1}} \cap \mathring{W}^1_{p_{i+1}}(\Omega), \quad Y_i = L^{p_i}(\Omega), \qquad i = 0,1,\ldots,N-1$$

and $X_0 = L^{q_0}(\Omega)$, where $p_i = \frac{q_i}{\alpha}$, $\frac{1}{q_{i+1}} = \frac{\alpha}{q_i} - \frac{2}{n}$, $i = 0,1,\ldots,N$; $q_0 = \frac{2n}{n-2}$ (for $n \geq 3$), with $0 < \alpha < \delta + \frac{(n+2)(1-\delta)}{n-2}$ for some $\delta > 0$.

We have, cf. K.C. Chang [23]:

THEOREM 3.3. *Suppose that* $f \in C^{2-0}(H,\mathbb{R}^1)$ *satisfies the P.S. condition.*

Assume that (3.6) *holds, and that the assumptions* (H_1)-(H_3) *are satisfied. Then the function* $\tilde{f}$ *possesses the deformation property.*

COROLLARY 3.2. *Under the assumptions in Theorem* 3.3, *we have*

$$C_*(\tilde{f},p_o) = C_*(f,p_o) \qquad \forall p_o \in K, \textit{ isolated.}$$

PROOF. For an open neighbourhood U of p_o, denote $V = \bigcup_{-\infty<t<+\infty} \eta(t,U)$, where η is the negative gradient flow. Then we have

$$\begin{aligned} C_*(f,p_o) &= H_*(f_c \cap V,(f_c \backslash \{p_o\}) \cap V;G), \qquad c = f(p_o) \\ &= H_*(\mathring{f}_{c+\varepsilon} \cap V, \mathring{f}_{c-\varepsilon} \cap V;G) \end{aligned}$$

for suitable $\varepsilon > 0$. Since now, V is open, applying (3.5) and Theorem 3.3, we obtain

$$C_*(f,p_o) = H_*(\mathring{\tilde{f}}_{c+\varepsilon} \cap V, \mathring{\tilde{f}}_{c-\varepsilon} \cap V;G) = C_*(\tilde{f},p_o). \ \Box$$

Section 4

CONNECTIONS WITH OTHER THEORIES

Besides Morse theory, there are two important topological methods in dealing with nonlinear problems. They are the degree theory and the Ljusternik-Schnirelman theory. In this section, we shall investigate the connections between Morse theory and them.

4.1. Degree theory

Let M and N be oriented n-dimensional manifolds without boundary and let

$$T:M \to N$$

be a smooth map. If M is compact and N is connected, then the Brouwer degree $\deg(T)$ is well defined [56].

Consider first an open set $U \subset \mathbb{R}^n$, and a smooth vector field

$$V:U \to \mathbb{R}^n$$

with isolated zero at the point $p \in U$. The function

$$\overline{V}(x) = \frac{V(x)}{\|V(x)\|}$$

maps a small sphere centered at p into the unit sphere S^{n-1}. The degree of $\overline{V}$

is called the index of V at the zero p, denoted by $\text{index}(V,p)$.

For a smooth vector field V on an arbitrary manifold M (finite dimensional) with isolated zero at the point $p \in M$ the index of V at the zero p is defined to be equal to the index of the corresponding vector field $dg^{-1} \circ V \circ g$ at the zero $g^{-1}(p)$, where $g:U \to M$ is a parametrization of a neighbourhood of p in M.

The celebrated Poincaré-Hopf theorem studies the relationship between the indices of zeros of a smooth vector field on M and the Euler characteristic of the manifold M. Namely

THEOREM (Poincaré-Hopf). *If* M^m *is a compact manifold with boundary* ∂M, *and if* V *is a smooth vector field on* M *which points outward at all boundary points, and has only isolated zeros, then the sum of the indices at the zeros* p_i, $i = 1,\ldots,n$, *of such a vector field is equal to the Euler characteristic of* M, *i.e.*,

$$\sum_{i=1}^{n} \text{index}(V,p_i) = \chi(M) \overset{\Delta}{=} \sum_{q=1}^{m} (-1)^q \text{ rank } H_q(M).$$

We shall extend this theorem to bounded domains on an infinite dimensional Hilbert space for a special class of vector fields, the gradient vector fields. Let H be a real Hilbert space, and let D be a bounded open set in H. For a vector field $V:\overline{D} \to H$, with $\theta \notin V(\partial D)$ and $\phi = id - V$ compact, the Leray-Schauder degree

$$\deg(V,D,\theta)$$

is well defined, cf. [60]. Suppose that p is an isolated zero of V, so there exists a ball $B_\varepsilon(p_o)$ with radius $\varepsilon > 0$ such that V has no zeros in $B_\varepsilon(p)$ other than p. Hence it is possible to define

$$\text{index}(V,p) = \deg(V,B_\varepsilon(p),\theta)$$

which is independent of ε for ε sufficiently small in view of the excision property of the Leray-Schauder degree. The number index(V,p) is called the index of V at the point p. In case $\dim H < +\infty$, the definition of index(V,p) coincides with the previous one.

We start with a local result in which the connection between the index of the gradient vector field of a function f at its isolated critical point and the critical groups of that point is studied; namely:

THEOREM 4.1. *Let* H *be a real Hilbert space, and let* $f \in C^2(H,\mathbb{R}^1)$ *be a function satisfying the P.S. condition. Assume that*

$$df(x) = x - T(x),$$

where T *is a compact mapping, and that* p_0 *is an isolated critical point of* f. *Then we have*

$$\text{ind}(df,p_0) = \sum_{q=0}^{\infty} (-1)^q \text{ rank } C_q(f,p_0). \tag{4.1}$$

PROOF. 1^0 Firstly, we assume that p_0 is nondegenerate. Since T is compact, we see that the Hessian

$$d^2f(p_0) = \text{id} - dT(p)$$

has only finite index j. By definition, and by Leray's formulas,

$$\text{index}(df,p_0) = (-1)^j.$$

In view of Theorem 1.1, we have

$$C_q(f,p_0) = \delta_{q,j}G.$$

Thus,

$$\text{index}(df,p_0) = \sum_{q=0}^{\infty} (-1)^q \text{ rank } C_q(f,p_0)$$

is proved in this special case.

2° For an isolated degenerate p_o we may assume for simplicity that $p_o = \theta$ and $f(p_o) = 0$. Let (W,W_-) be a Gromoll-Meyer pair constructed in (2.11) and (2.12), and let $\delta > 0$ be sufficiently small such that $B_\delta \subset \mathring{W} \cap f^{-1}[-\frac{\gamma}{3},\frac{\gamma}{3}]$, where γ is the real number appearing in (2.11).

We shall define a function $\tilde{f}$ satisfying the P.S. condition such that

(1) $|f(x) - \tilde{f}(x)| < \frac{\gamma}{3}$ $\forall x \in H$.

(2) $df(x) = d\tilde{f}(x)$ for x in a neighbourhood of ∂W.

(3) In W, $\tilde{f}$ has only nondegenerate critical points $\{p_j\}_1^m$, finite in number, contained in B_δ.

Once the function $\tilde{f}$ is constructed, we obtain immediately

$$W_- = f_{-\gamma} \cap W \subset \tilde{f}_{-\frac{2}{3}\gamma} \cap W \subset f_{-\frac{\gamma}{3}} \cap W \subset f_{\frac{\gamma}{3}} \cap W \subset \tilde{f}_{\frac{2}{3}\gamma} \cap W \subset f_\gamma \cap W = W.$$

However, there are strong deformation retracts:

$$f_\gamma \cap W \to f_{\frac{\gamma}{3}} \cap W \quad \text{and} \quad f_{-\frac{\gamma}{3}} \cap W \to f_{-\gamma} \cap W$$

provided by the Gromoll-Meyer property. We have

$$H_*(W,W_-) = H_*\left(\tilde{f}_{\frac{2}{3}\gamma} \cap W,\ \tilde{f}_{-\frac{2}{3}\gamma} \cap W\right) \tag{4.2}$$

due to the exactness of the homological group sequence. Thus,

$$\begin{aligned}
\text{index}(df,\theta) &= \deg(df,W,\theta) && \text{(by (2.14), (2.15))}\\
&= \deg(d\tilde{f},W,\theta) && \text{(by (2))}\\
&= \sum_{j=1}^{m} \text{index}(d\tilde{f},p_j) && \text{(by (3))}\\
&= \sum_{j=1}^{m} \sum_{q=0}^{\infty} (-1)^q \operatorname{rank} C_q(\tilde{f},p_j) && \text{(by } 1^o) .
\end{aligned}$$

Applying Theorem 1.4 and Remark 1.3 to the function $\tilde{f}$ on W, we have

$$\sum_{j=1}^{m} \sum_{q=0}^{\infty} (-1)^q \operatorname{rank} C_q(\tilde{f},p_j) = \sum_{q=0}^{\infty} (-1)^q \operatorname{rank} H_q\left(\tilde{f}_{\frac{2}{3}\gamma} \cap W, \tilde{f}_{-\frac{2}{3}\gamma} \cap W\right)$$

$$= \sum_{q=0}^{\infty} (-1)^q \operatorname{rank} H_q(W,W_-)$$

$$= \sum_{q=0}^{\infty} (-1)^q \operatorname{rank} C_q(f,\theta).$$

This is due to the fact that the negative gradient flow of $\tilde{f}$ directs inward on $\partial W \backslash W_-$, and hence also on $\partial W \backslash (W \cap \tilde{f}^{-1}(-\frac{2}{3}\gamma))$.

3° Finally, we shall construct a function $\tilde{f}$, satisfying the P.S. condition as well as the conditions (1)-(3). We define

$$\tilde{f}(x) = f(x) + p(\|x\|)(x_o,x)$$

where $p \in C^2(\mathbb{R}^1_+,\mathbb{R}^1)$ is a function satisfying

$$p(t) = \begin{cases} 1 & 0 \le t \le \delta/2 \\ 0 & t > \delta \end{cases}$$

with $0 \le p \le 1$ and $|p'(t)| \le \frac{4}{\delta}$, $x_o \in H$ is determined later. Let

$$\beta = \inf\{\|df(x)\| \mid x \in B_\delta \backslash B_{\delta/2}\}$$

then $\beta > 0$. We choose $x_o \in H$ such that

$$0 < \|x_o\| < \min\{\frac{\beta}{6},\frac{\gamma}{3}\}.$$

Then we have

$$|f(x) - \tilde{f}(x)| < \frac{\gamma}{3} \qquad \forall x \in H$$

$$\|d\tilde{f}(x)\| \ge \frac{\beta}{6} \qquad \forall x \in B_\delta \backslash B_{\delta/2}$$

$$d\tilde{f}(x) = df(x) \qquad \forall x \notin B_\delta.$$

For smaller $\|x_o\|$, $d\tilde{f}$ is a k-set contraction mapping vector field with $k < 1$ (cf. Lloyd [52]). Therefore $\deg(d\tilde{f},W,\theta)$ is well defined. The P.S. condition for the function $\tilde{f}$ is verified directly. On the account of the Sard-Smale theorem, a suitable x_0 can be chosen such that $\tilde{f}$ is nondegenerate. The proof is completed.

We generalize Theorem 4.1 as follows:

THEOREM 4.2. *Under the assumption in Theorem 4.1, suppose that* W *is a bounded domain in* H *on which* f *is bounded. Assume that*

(1) $W_- \overset{\Delta}{=} \{x \in \partial W \mid \eta(t,x) \notin W \ \ \forall t > 0\} = W \cap f^{-1}(a)$ *for some* a, *where* $\eta(t,x)$ *is the negative gradient flow of* f *emanating from* x;

(2) $-df|_{\partial W \setminus W_-}$ *directs inward.*

Then we have

$$\deg(df,W,\theta) = \chi(W,W_-). \tag{4.3}$$

PROOF. Due to the assumptions (1) and (2), $\theta \notin df(\partial W)$, the Leray-Schauder degree $\deg(df,W,\theta)$ is well defined.

If f is nondegenerate on W, then the critical set K consists of finitely many isolated points $\{p_j\}_1^m$, because f is bounded on W, and the assumption (2), as well as the P.S. condition holds. According to Theorem 4.1, we have

$$\begin{aligned}\deg(df,W,\theta) &= \sum_{j=1}^{m} \text{index}(df,p_j) \\ &= \sum_{j=1}^{m} \sum_{q=0}^{\infty} (-1)^q \text{ rank } C_q(f,p_j) \\ &= \sum_{q=0}^{\infty} (-1)^q \text{ rank } H_q(W,W_-).\end{aligned}$$

The last equality follows from the assumption (2), Theorem 1.4 (Remark 1.3) and the fact that $f(p_j) > a$, which is a consequence of the assumption (1).

If f is degenerate, we perturbe it as in Theorem 4.1. Since the critical set K is compact in $\mathring{W}$, we construct a C^2-function $p(x)$, satisfying (cf. Bonic-Frampton [17]):

$$p(x) = \begin{cases} 0 & x \in \Omega_\delta, \\ 1 & x \notin \Omega_{2\delta}, \end{cases}$$

where $\delta > 0$ such that $\operatorname{dist}(K, \partial W) \geq 2\delta$, K is the critical set of f in W, and

$$\Omega_\delta = \{x \in W \mid \operatorname{dist}(x, \partial W) < \delta\}.$$

We may assume that

$$|p(x)| \leq 1 \quad \text{and} \quad |p'(x)| \leq M < +\infty.$$

Let

$$M_1 = \sup\{\|x\| \mid x \in W\}, \qquad b = \inf\{f(x) \mid x \in W \backslash \Omega_\delta\}$$

and

$$\beta = \min\{\inf_{x \in \Omega_{2\delta}} \|df(x)\|, 1\}.$$

Then, by our assumptions, $b > a$ and $\beta > 0$. One defines

$$\tilde{f}(x) = f(x) + p(x)(x_0, x)$$

for suitable $x_0 \in H$, with

$$0 < \|x_0\| < \min\{\frac{b-a}{M_1}, \frac{\beta}{3}, \frac{\beta}{3M_1M_2}\}.$$

By the Sard-Smale theorem, x_0 can be chosen such that $\tilde{f}$ is nondegenerate, because

$$\begin{aligned} \|d\tilde{f}(x)\| &\geq \|df(x)\| - \|x_0\| - \|p'(x)\| \|x\| \cdot \|x_0\| \\ &\geq \beta - \frac{1}{3}\beta - M_1 M_2 \|x_0\| \\ &> \frac{1}{3}\beta > 0 \qquad \forall x \in \Omega_{2\delta}. \end{aligned}$$

Now the function $\tilde{f}$ satisfies the P.S. condition and the assumptions (1) and (2). In fact, the P.S. condition is easily verified by the above estimate, assumption (2) is trivial, and the assumption (1) is verified by the following inequality:

$$\tilde{f}(x) \geq f(x) - \|x_0\|\|x\| \geq f(x) - \frac{b-a}{M_1} \cdot M_1 > a \qquad \forall x \in W\backslash\Omega_\delta .$$

Since

$$\deg(df,W,\theta) = \deg(d\tilde{f},W,\theta)$$

(the RHS is the generalized Leray-Schauder degree for a k-set contraction mapping vector field with $k < 1$), the proof is completed.

REMARK 4.1. Theorem 4.1, which has been given by E. Rothe [71,72] under stronger assumptions, was obtained by Hofer [42] and Tian [79] by a combination of the Poincaré-Hopf theorem with the splitting theorem.

4.2. Ljusternik-Schnirelman theory

Let M be a C^2-Finsler manifold. The category of M is defined by

$$\operatorname{cat}(M) = \inf\{m \in \mathbb{N} \cup \{+\infty\} \mid \exists \text{ contractible closed subsets } F_1,\ldots,F_m \text{ such that } M \subset \bigcup_{i=1}^{m} F_i\}.$$

We have the following

(Ljusternik-Schnirelman theorem) *Suppose that* $f \in C^1(M,\mathbb{R}^1)$ *is a function bounded from below, satisfying the P.S. condition. Then* f *has at least* $\operatorname{cat}(M)$ *critical points.*

It is known [75] that the topological invariant $\operatorname{cat}(M)$ can be estimated by other topological invariants. For instance, let $H^*(M)$ be the cohomology ring of M, with cup product $\cup$; one defines the cuplength of M to be equal to

$$\sup\{m \in N \mid \exists x_1,\ldots,x_m \in H^*(M) \text{ such that } \dim x_i > 0, \; i = 1,\ldots,m, \text{ and } x_1 \cup \ldots \cup x_m \neq 0\}.$$

The cuplength of M provides a lower bound for $\operatorname{cat}(M)$, i.e.,

$$\operatorname{cat}(M) \geq (\text{cuplength of } M) + 1. \tag{4.4}$$

Now we extend this relation to estimate the number of critical points by relative homology groups.

Recall that the cap product is a dual operation of the cup product: $\forall p$-cochain c, $\forall q$-cochain d, and $(p+q)$-chain σ,

$$(\sigma \cap c, d) = (\sigma, c \cup d),$$

where $(\ ,\ )$ is the duality between chains and cochains. We shall use the cap product in replacing the cup product in this extension.

DEFINITION 4.1. Let (X,Y) be a pair of topological spaces with $Y \subset X$; for two nontrivial singular homology classes $[z_1],[z_2] \in H_*(X,Y)$ we define $[z_1] < [z_2]$, if $\exists \omega \in H^*(X)$ with $\dim \omega > 0$ such that $[z_1] = [z_2] \cap \omega$.

LEMMA 4.1. *Suppose that* $f \in C^1(M,\mathbb{R}^1)$ *satisfies the P.S. condition. Let* $a < b$ *be two real numbers such that* f *has only isolated critical points in* $f^{-1}[a,b]$. *Assume that* $[z_1],[z_2] \in H_*(f_b,f_a;G)$ *with* $[z_1] < [z_2]$. *Then*

$$c_i = \inf_{\tilde{z}_i \in [z_i]} \sup_{x \in |\tilde{z}_i|} f(x), \qquad i = 1,2,$$

are critical values of f *with* $a < c_1 \le c_2 \le b$, *where* $|\tilde{z}|$ *denotes the support of the singular chain* $\tilde{z}$.

PROOF. According to the Minimax Principle, c_1, c_2 are critical values with $a < c_1 \le c_2 \le b$. It remains to prove $c_1 < c_2$.

Suppose that $c_1 = c_2 = c$, then $\forall \varepsilon > 0$ we have $\tilde{z}_2 \in [z_2]$ with $|\tilde{z}_2| \subset f_{c+\varepsilon}$. Since K_c consists of isolated critical points, we may choose two contractible neighbourhoods of K_c, $N \subset N'$. There is a cochain $\tilde{\omega} \in \omega$ with support in $f_b \backslash N'$, provided by $\dim \omega > 0$ and that N' is contractible. Subdividing $\tilde{z}_2$, $\tilde{z}_2 = \tilde{z}_2' + \tilde{z}_2''$ such that $|\tilde{z}_2'| \subset N'$, and $|\tilde{z}_2''| \subset f_{c+\varepsilon} \backslash N$, we have

$$|\tilde{z}_1| \subset f_{c+\varepsilon} \backslash N$$

because $\tilde{z}_1 = \tilde{z}_2 \cap \tilde{\omega} = \tilde{z}_2'' \cap \tilde{\omega}$.

According to the deformation lemma, there is a homeomorphism $\eta: f_{c+\varepsilon} \backslash N \to f_{c-\varepsilon}$. This implies $\eta|\tilde{z}_1| \subset f_{c-\varepsilon}$. But $\eta(\tilde{z}_1) \in [z_1]$. This contradicts the definition of c_1 (= c). □

THEOREM 4.3. *Suppose that* $f \in C^1(M,R^1)$ *satisfies the P.S. condition. Let* $a < b$ *be two real numbers such that* f *has only isolated critical points in* $f^{-1}[a,b]$. *Assume that there are* m *nontrivial singular homology classes* $[z_1],[z_2],\ldots,[z_m] \in H_*(f_b,f_a;G)$ *with* $[z_1] < [z_2] < \ldots < [z_m]$. *Then* f *has at least* m *distinct critical values.*

PROOF. Straightforward.

As a corollary, we return to the following result which is a combination of the Ljusternik-Schnirelman theorem with (4.4).

COROLLARY 4.1. *Suppose that* $f \in C^1(M,\mathbb{R}^1)$, *satisfying the P.S. condition, is bounded from below. Then* f *has at least* (cuplength of M) + 1 *distinct critical points.*

PROOF. By definition, $\exists \omega_1,\ldots,\omega_m \in H^*(M)$ with $\dim \omega_i > 0$, $i = 1,\ldots,m$, such that $\omega_1 \cup \ldots \cup \omega_m \neq 0$, where m = cuplength of M. There exists $[z_1] \in H_*(M)$ such that

$$<[z_1],\omega_1 \cup \ldots \cup \omega_m> \neq 0$$

where $<\ ,\ >$ denotes the duality between $H_*(M)$ and $H^*(M)$. Define

$$[z_{i+1}] = [z_i] \cap \omega_i, \qquad i = 1,\ldots,m.$$

We obtain m+1 nontrivial singular homology classes $[z_{m+1}] < [z_m] < \ldots < [z_1]$ in $H_*(M)$. If we take $a = \inf\{f(x) \mid x \in M\}$ and b large enough such that $[z_1],\ldots,[z_{m+1}]$ are nontrivial in f_b, the conclusion follows from Theorem 4.3.

Section 5

THREE CRITICAL POINT THEOREM

In this section we apply the Morse identity (1.7) to estimate the number of critical points for a class of functions. Applications to PDE are provided as examples.

THEOREM 5.1. *Suppose that* M *is a* C^2 *Finsler manifold. Assume that* $f \in C^1(M,\mathbb{R}^1)$, *satisfying the P.S. condition, is bounded below. Suppose that there is a critical point* p_o, *which is not the global minimum of* f, *with finite* $\sum_{q=0}^{\infty} (-1)^q \operatorname{rank} C_q(f,p_o) \neq \chi(M) - 1$. *Then* f *has at least three critical points.*

PROOF. By the P.S. condition and the lower semi-boundedness, f has a global minimum p_1. Let $c_i = f(p_i)$, $i = 0,1$. If f had no critical point other than p_o and p_1, then for arbitrary $b > c_o$, there would be no critical point in $M\backslash f_b$, and the following identity would hold:

$$\chi(f_b) = \chi(f_b, f_{c_o-\varepsilon}) + \chi(f_{c_o-\varepsilon}),$$

where $0 < \varepsilon < c_o - c_1$. Since there exists a strong deformation retract deforming M into f_b, and $f_{c_o-\varepsilon}$ into p_1; we would have

$$\chi(f_b) = \chi(M)$$

and

$$\chi(f_{c_0-\varepsilon}) = \chi(\{p_1\}) = 1.$$

But

$$\chi(f_b, f_{c_0-\varepsilon}) = \sum_{q=0}^{\infty} (-1)^q \text{ rank } C_q(f, p_0)$$

because p_0 is the unique critical point in $f^{-1}[c_0-\varepsilon, b]$. This is a contradiction.

The theorem has several corollaries very useful in applications.

COROLLARY 5.1. *Suppose that* M *is a* C^2 *Finsler manifold with an equivalent differentiable norm, and that* $\chi(M)$ *is odd. Assume that* $f \in C^1(M, \mathbb{R}^1)$ *is bounded below, satisfies the P.S. condition, is* C^2 *in a neighbourhood* U *of its nondegenerate critical point* p_0 *with finite index, and that* p_0 *is not the global minimum. Then* f *has at least three critical points.*

PROOF. Applying Theorem 5.1 and Corollary 3.1 we obtain the conclusion directly.

COROLLARY 5.2. *Let* H *be a Hilbert space, and* $f \in C^2(H, \mathbb{R}^1)$ *be bounded below with the P.S. condition. Suppose that* $df(x) = x - T(x)$ *is a compact vector field, and that* p_0 *is an isolated critical point but not the global minimum with* $\text{index}(df, p_0) = \pm 1$. *Then* f *has at least three critical points.*

PROOF. Applying Theorems 5.1 and 4.1, the conclusion follows.

COROLLARY 5.3. *Let* $f \in C^1(\mathbb{R}^n, \mathbb{R}^1)$ *be bounded from below, with the P.S. condition. Assume that* f *has a local maximum. Then* f *has at least three critical points.*

PROOF. Straightforward from Theorem 5.1 and Example 1 in Section 1.

REMARK 5.1. In Castro and Lazer [20], Theorem 5.1 was proved in a much more roundabout way under the assumptions that $M = \mathbb{R}^n$, and $f(x) \to +\infty$ as $\|x\| \to +\infty$. Theorem 5.1 was extended to Hilbert space in Chang [21]. Amann [3] pointed out that for the finite dimensional case, Castro and Lazer's result had been already obtained earlier by Krasnoselskii [47] via degree theory. Along the lines of Krasnoselskii, Amann gave a proof for f with a gradient df as a compact vector field, defined on a Hilbert space, i.e., Corollary 5.2.

Theorem 5.1, as well as its corollaries, will be applied to different problems in PDE later. In this section we shall present an application in which a Banach space setting is needed.

Consider the following boundary value problem

$$c \sum_{i=1}^{n} \frac{\partial}{\partial x_i}\left(\left|\frac{\partial u}{\partial x_i}\right|^{p-2} \frac{\partial}{\partial x_i} u\right) + \Delta u = g(u) \quad \text{in } \Omega \tag{5.1}$$

$$u|_{\partial\Omega} = 0$$

where Ω is a bounded domain in $\mathbb{R}^n$ with smooth boundary $\partial\Omega$, $p > 2$, and $c > 0$ is a constant. The function g satisfies the following assumptions:

(1) $g \in C^1(\mathbb{R}^1)$ with $g(0) = 0$.

(2) $|g'(t)| \le \begin{cases} C(1 + |t|^{\alpha-1}) & \text{if } p \le n \\ \text{no restriction in growth, if } p > n \end{cases}$

where

$$\alpha < \frac{np}{n-p} - 1.$$

(3) Let $G(\xi) = \int_0^\xi g(t)\, dt$. We assume that

$$G(\xi) \ge -c_o|u|^p - c_1$$

with

$$c_o < \frac{c}{p} \inf\{\int_\Omega \sum_{i=1}^{n} |\partial_i u|^p / \int_\Omega |u|^p \mid u \in C_o^\infty(\Omega)\}.$$

(4) $\exists i \ge 1$ such that

$$\lambda_i < -g'(0) < \lambda_{i+1}$$

where $\{\lambda_j\}_1^\infty = \sigma(-\Delta)$, the spectrum of the Laplacian with 0-Dirichlet data.

THEOREM 5.2. *The equation* (5.1) *has at least three solutions.*

PROOF. It is carried out by considering the following functional

$$f(u) = \frac{c}{p}\int_\Omega \sum_{i=1}^n |\partial_i u|^p + \tfrac{1}{2}\int_\Omega |\nabla u|^2 + \int_\Omega G(u) \tag{5.2}$$

on the Sobolev space $\overset{\circ}{W}{}^1_p(\Omega)$.

1° To verify $f \in C^2$, we need

LEMMA 5.1. $\forall \varepsilon \in 0\ \exists \delta \in 0$ *such that*

$$\left|\int_\Omega [g(u) - g'(0)u]v\right| \le \varepsilon\left(\|u\|_{\overset{\circ}{W}{}^1_2} \cdot \|v\|_{\overset{\circ}{W}{}^1_2} + \|u\|^{p-1}_{\overset{\circ}{W}{}^1_p} \cdot \|v\|_{\overset{\circ}{W}{}^1_p}\right)$$

$\forall u,v \in \overset{\circ}{W}{}^1_p(\Omega)$, *with* $\|u\|_{W^1_p} < \delta$.

Thus

$$\langle df(u),v\rangle = \int_\Omega \sum_{i=1}^n (c|\partial_i u|^{p-2} + 1)\partial_i u \partial_i v + g(u)v$$

and

$$d^2 f(v,w) = \int_\Omega \sum_{i=1}^n (c(p-1)|\partial_i u|^{p-2} + 1)\partial_i v \partial_i w + g'(u)vw.$$

2° The assumption (3) implies the lower semi-boundedness of f.

3° The P.S. condition.

Let $G: u \to g(u)$ be the Nemytcki operator. Due to the assumption (2),

$$\overset{\circ}{W}{}^1_p \hookrightarrow L^{q_1} \overset{G}{\to} L^{q_1'} \to W^{-1}_{p'} \to W^1_{p'}$$

where $q_1 < p$, $\alpha = \frac{q_1}{q_1'}$, $\frac{1}{q_1} + \frac{1}{q_1'} = 1$ and $\frac{1}{p} + \frac{1}{p'} = 1$.

Since $K = (-\Delta)^{-1} \in L(W^{-1}_r, \overset{\circ}{W}{}^1_r)$, K is compact in $L(\overset{\circ}{W}{}^1_r, W^1_r)$ for $1 < r < \infty$. Let

$$R(u) = cK\sum_{i=1}^n \partial_i(|\partial_i u|^{p-2}\partial_i u) + u$$

we have $R:\overset{\circ}{W}{}^1_p \to \overset{\circ}{W}{}^1_{p'}$ continuously, and

$$\langle df(u),v\rangle = \langle Ru + KG(u),v\rangle .$$

Suppose that $\{u_n\}$ is a sequence in $\overset{\circ}{W}{}^1_p$ such that

$$f(u_n) \text{ is bounded and}$$
$$R(u_n) + KG(u_n) \to \theta \qquad (\overset{\circ}{W}{}^1_{p'}).$$

Then $\{u_n\}$ is bounded in $\overset{\circ}{W}{}^1_p$, provided by

$$f(u) \geq \tfrac{1}{2}\int_\Omega (\nabla u)^2 - C_1 \mathrm{mes}(\Omega) + \left(\frac{c}{p} - \frac{c_o}{\bar{c}_p}\right)\int_\Omega \sum_{i=1}^n |\partial_i u|^p$$

where

$$\bar{c}_p = \inf\{\int_\Omega \sum |\partial_i u|^p / \int_\Omega |u|^p \mid u \in C_o^\infty(\Omega)\}.$$

We have a subsequence $\{u_{n_i}\}$ such that $KG(u_{n_i})$ is strongly convergent in $\overset{\circ}{W}{}^1_{p'}$, say to w^*. This implies that

$$R(u_{n_i}) \to -w^* \qquad (\overset{\circ}{W}{}^1_{p'}).$$

However, the operator R is strongly monotone; we conclude that $\{u_{n_i}\}$ converges strongly in $\overset{\circ}{W}{}^1_p$.

4° Now we turn to verify the nondegeneracy of the critical point θ, and to compute the index at θ. A hyperbolic operator

$$L = id + g'(0)(-\Delta)^{-1} \in L(\overset{\circ}{W}{}^1_p(\Omega),\overset{\circ}{W}{}^1_p(\Omega))$$

is defined, $\sigma(L) = \{1 + g'(0)/\lambda_j \mid j \in \mathbb{N}^*\}$. By definition,

$$d^2f(\theta)(Lu,u) = \int_\Omega \nabla(Lu)\nabla u + g'(0)Lu\cdot u$$
$$= \int_\Omega |\nabla(Lu)|^2 \geq \lambda_1 \int_\Omega (Lu)^2 > 0 \qquad \text{if } u \neq \theta,$$
$$d^2f(\theta)(Lu,v) = \int_\Omega \nabla(Lu)\cdot\nabla(Lv) = d^2f(\theta)(u,Lv) \qquad \forall u,v \in \overset{\circ}{W}{}^1_p,$$

and

$$\text{index } d^2 f(\theta) = i.$$

To verify

$$(5.3) \qquad \langle df(u), Lu\rangle = \int_\Omega \left[\sum_{i=1}^n (c|\partial_i u|^{p-2} + 1)\partial_i u \partial_i (Lu) + g(u)Lu\right] > 0$$

for u, $0 < \|u\|_{\mathring{W}_p^1} < \varepsilon$. We need the following

LEMMA 5.2. *There exist constants* $\beta = \beta(n,p) \in (0,1]$ *and* $M = M(n,p) > 0$ *such that*

$$\frac{\int_\Omega |u|^p}{\int_\Omega \sum_{i=1}^n |\partial_i u|^p} \le M \left(\frac{\int_\Omega |u|^2}{\int_\Omega |\nabla u|^2}\right)^{\beta p/2}, \qquad \forall u \in \mathring{W}_p^1 \backslash \{\theta\}.$$

This is the Gagliardo-Nirenberg inequality. From it we obtain

LEMMA 5.3. *Let*

$$J(u) = \int_\Omega \sum_{i=1}^n |\partial_i u|^{p-1} |\partial_i K u|.$$

Then $\forall \varepsilon > 0$ $\exists N$, *a positive integer, such that*

$$J((id - P_m)u) \le \varepsilon \int_\Omega \sum_{i=1}^n |\partial_i (id - P_m)u|^p$$

for $m > N$, *where* P_m *is the restriction on* $\mathring{W}_p^1$ *of the orthogonal projection onto* $E_k = \text{span}\{\phi_1, \ldots, \phi_k\}$ *in* $L^2(\Omega)$, *and* ϕ_k *is the eigenfunction corresponding to* λ_k, $k = 1,2,\ldots$.

Now we turn to (5.3).

$$\begin{aligned}\langle df(u), Lu\rangle &= \int_\Omega |\nabla(Lu)|^2 + c\int_\Omega \sum |\partial_i u|^{p-2}\partial_i u \partial_i (Lu) + \int_\Omega (g(u) - g'(0)u)Lu \\ &\ge \nu \int_\Omega (\nabla u)^2 + c\int_\Omega \sum_{i=1}^n |\partial_i u|^p - c|g'(0)|J(u) - \varepsilon\left(\|u\|^2_{W_2^1} + \|u\|^p_{W_p^1}\right)\end{aligned}$$

as $\|u\|_{\mathring{W}_p^1} < \delta$, provided by Lemma 5.2, where

$$\nu = \mathrm{Min}\left(1 + \frac{g'(0)}{\lambda_{i+1}}\ ,\ -\left(1 + \frac{g'(0)}{\lambda_i}\right)\right),$$

and $\varepsilon > 0$ is chosen $< \min\{\frac{\nu}{2}, \frac{c}{2}\}$. Hence

$$\langle df(u), Lu\rangle \geq \frac{\nu}{2}\|u\|^2_{\overset{\circ}{W}{}^1_2} + \frac{c}{2}\|u\|^p_{\overset{\circ}{W}{}^1_p} - c|g'(0)|J(u).$$

However

$$J(u) \leq 2^{p-1}\int_\Omega \sum_{i=1}^n (|\partial_i P_N u|^{p-1} + |\partial_i(id-P_N)u|^{p-1})(|\partial_i KP_N u| + |\partial_i K(id-P_N)u|)$$

$$\leq 2^{p-1} J((id-P_N)u) + R_N(u)$$

where $R_N(u)$ is the remainder. There exists a constant C_N, depending on N, such that

$$R_N(u) \leq C_N\|P_N u\|^p_{\overset{\circ}{W}{}^1_p} .$$

Noticing that

$$\|u\|^p_{\overset{\circ}{W}{}^1_p} \geq \frac{1}{2^p}\|(id-P_N)u\|^p_{\overset{\circ}{W}{}^1_p} - \|P_N u\|^p_{\overset{\circ}{W}{}^1_p}$$

we arrive at

$$\langle df(u), Lu\rangle \geq \frac{\nu}{2}\|u\|^2_{\overset{\circ}{W}{}^1_2} + \{\frac{c}{2^{p+1}}\|(id-P_N)u\|^p_{\overset{\circ}{W}{}^1_p} - 2^p c|g'(0)|J((id-P_N)u)\} - C'_N\|P_N u\|^p_{\overset{\circ}{W}{}^1_p},$$

where C'_N denotes a constant depending on N. Applying Lemma 5.3, N is chosen such that the bracket is positive. Fixing N, the norms $\|.\|_{\overset{\circ}{W}{}^1_2}$ and $\|.\|_{\overset{\circ}{W}{}^1_p}$ are equivalent in the finite dimensional space $P_N\overset{\circ}{W}{}^1_p$, then the inequality (5.3) is proved for $p > 2$.

REMARK 5.2. The same conclusion holds true for $p = 2$. In this case the proof is straightforward, because it works in a Hilbert space $(c = 0)$, cf. K.C. Chang [21].

REMARK 5.3. The proofs of Lemmas 5.1-5.3 can be found in Chang [22].

REMARK 5.4. Other applications to different kinds of differential equations can be found in Chang [21], Chang, Wu, Li [27], Castro and Lazer [20].

Section 6

HOMOTOPY INVARIANCE AND APPLICATIONS TO BIFURCATION PROBLEMS

The critical groups and the Leray-Schauder degree are all topological characterizations of the critical points. Theorem 4.1 provides a relationship between them. Since $\text{index}(df,p_o) = \sum_{q=0}^{\infty} (-1)^q \text{ rank } C_q(f,p_o)$, we see that the critical groups provide more information than the degree. In this aspect, the Morse theory has the advantage of the degree theory. On the other hand, the degree is homotopy invariant, i.e., it is stable under small perturbation; but the critical groups are not, because under a small perturbation, an isolated critical point might be changed to a non-isolated one. Generally speaking, we have only the lower semi-continuity of the Morse type numbers.

LEMMA 6.1. *Let* $A \subset X \subset B \subset A' \subset Y \subset B'$ *be topological spaces. Suppose that* $H_*(B,A) \cong H_*(B',A') \cong 0$. *Then* $h: H_*(A',A) \to H_*(Y,X)$ *is an injection.*

PROOF. Observing the following diagrams:

$$\begin{array}{ccccccc} H_{q+1}(B',A') & \longrightarrow & H_q(A',A) & \xrightarrow{i} & H_q(B',A) & \longrightarrow & H_q(B',A') \\ & & & \searrow^{\alpha} \quad \nearrow^{\beta} & & & \\ & & & H_q(Y,A) & & & \end{array},$$

$$\begin{array}{ccccccc} H_q(B,A) & \longrightarrow & H_q(Y,A) & \xrightarrow{i_1} & H_q(Y,B) & \longrightarrow & H_{q-1}(B,A) \\ & & & \searrow^{\alpha_1} \quad \nearrow^{\beta_1} & & & \\ & & & H_q(Y,X) & & & \end{array} ,$$

i, i_1 are isomorphisms. However $i = \beta \circ \alpha$, $i_1 = \beta_1 \circ \alpha_1$. Therefore α_1 and α are injections, and $h = \alpha_1 \circ \alpha$ is an injection.

THEOREM 6.1. *Suppose that* $f \in C^2(H,\mathbb{R}^1)$ *satisfies the P.S. condition, with an isolated critical value* c, *where* H *is a Hilbert space. Assume that* (a,b) *is an interval containing* c. *Then there exists an* $\varepsilon > 0$ *such that for*

$$\|g-f\|_{C(f^{-1}[a,b])} < \frac{\varepsilon}{3}, \tag{6.1}$$

we have an injection

$$i:H_*(f_{c+\varepsilon}, f_{c-\varepsilon}) \to H_*\left(g_{c+\frac{\varepsilon}{2}}, g_{c-\frac{\varepsilon}{2}}\right).$$

PROOF. Choose $\varepsilon > 0$ such that c is the only critical value of f in $[c-\varepsilon, c+\varepsilon] \subset (a,b)$. (6.1) implies

$$f_{c-\varepsilon} \subset g_{c-\frac{\varepsilon}{2}} \subset f_{c-\frac{\varepsilon}{6}} \subset f_{c+\frac{\varepsilon}{6}} \subset g_{c+\frac{\varepsilon}{2}} \subset f_{c+\varepsilon}.$$

Applying Lemma 6.1, we obtain

$$i:H_*\left(f_{c+\frac{\varepsilon}{6}}, f_{c-\varepsilon}\right) \to H_*\left(g_{c+\frac{\varepsilon}{2}}, g_{c-\frac{\varepsilon}{2}}\right).$$

is an injection. Hence $H_*(f_{c+\varepsilon}, f_{c-\varepsilon}) \to H_*\left(g_{c+\frac{\varepsilon}{2}}, g_{c-\frac{\varepsilon}{2}}\right)$. is an injection.

THEOREM 6.2. *Suppose that* $f \in C^1(H,\mathbb{R}^1)$, *satisfying the P.S. condition, has only finitely many critical points in* $f^{-1}[a,b]$, *where* a,b *are regular values of* f. *Then there exists an* $\varepsilon > 0$ *such that*

$$M_q(f) \le M_q(g), \qquad q = 0,1,2,\ldots$$

for g, $\|g-f\|_{C(f^{-1}[a,b])} < \varepsilon$, *where* $M_q(\ )$ *are the Morse type numbers w.r.t.* (a,b), *and* $g \in C^1(H,\mathbb{R}^1)$ *with the P.S. condition.*

PROOF. Straightforward.

Theorem 6.2 means that the Morse type numbers are lower semi-continuous under C^o-perturbation.

Next, we turn to the continuity property of the critical groups under stronger restrictions.

LEMMA 6.2. *Let* (W,W_-) *be a Gromoll-Meyer pair of an isolated critical point* p_o *of a* C^1 *function* f, *defined on a Hilbert space* H *satisfying the P.S. condition, w.r.t. the negative gradient vector field* $-df(x)$. *Then there exists* $\varepsilon > 0$ *such that* (W,W_-) *is also a Gromoll-Meyer pair of* g *satisfying the P.S. condition w.r.t. certain p.g.v.f. of* g, *provided that* p_o *is the only critical point of* g *in* W, *and* $\|g-f\|_{C^1(\overline{W})} < \varepsilon$.

PROOF. $\exists r > 0$ such that the ball $B(p_o,r) \subset \mathring{W}$. Due to the P.S. condition

$$\beta = \inf\{\|df(x)\| \mid x \in W\backslash B(p_o,r/2)\} > 0.$$

Define a function $\rho(x) \in C^1(H,\mathbb{R}^1)$, satisfying

$$\rho(x) = \begin{cases} 1, & x \in B(p_o,r/2) \\ 0, & x \notin B(p_o,r) \end{cases}$$

with $0 \le \rho(x) \le 1$, and a vector field

$$V(x) = \frac{3}{2}[\rho(x)dg(x) + (1-\rho(x))df(x)].$$

Choosing $0 < \varepsilon < \beta/4$ we obtain

$$\left.\begin{aligned} &\|V(x)\| \le 2\|dg(x)\| \\ &(V(x),dg(x)) \ge \|dg(x)\|^2 \end{aligned}\right. \quad \forall x \in \overline{W},$$

for $\|g-f\|_{C^1(\overline{W})} < \varepsilon$; because for $x \notin B(p_o,r/2)$, $\|dg(x)\| \geq \|df(x)\| - \varepsilon \geq \beta - \varepsilon \geq 3\varepsilon$, we have

$$(V(x),dg(x)) \geq \frac{3}{2}[\|dg(x)\|^2 - \varepsilon\|dg(x)\|]$$
$$\geq \frac{3}{2}[\|dg(x)\|^2 - \frac{1}{3}\|dg(x)\|^2] = \|dg(x)\|^2;$$

and for $x \in B(p_o,r/2)$, one has trivially:

$$(V(x),dg(x)) = \frac{3}{2}\|dg(x)\|^2 \geq \|dg(x)\|^2.$$

Since $V(x) = -df(x)$ outside a ball $B(p_o,r) \subset \overset{\circ}{W}$, and p_o is the only critical point of g in W, the conclusion follows.

THEOREM 6.3. *Suppose that* $f \in C^2(H,\mathbb{R}^1)$ *satisfies the P.S. condition with unique critical point* p_o *in a ball* $B(p_o,\delta)$, $\delta > 0$. *Then there exists an* $\varepsilon > 0$ *such that*

$$C_*(f,p_o) = C_*(g,p_o)$$

for $g: \|f-g\|_{C^1(B(p_o,\delta))} < \varepsilon$, *which preserves* p_o *as the only critical point in* $B(p_o,\delta)$, *with the P.S. condition.*

PROOF. According to Theorem 2.3, we may construct a Gromoll-Meyer pair (W,W_-) w.r. to $-df(x)$ with $W \subset B(p_o,\delta)$. Taking the same $\varepsilon > 0$ as in Lemma 6.2, we see that (W,W_-) is a Gromoll-Meyer pair w.r.t. a certain p.g.v.f. of g, which satisfies $\|f-g\|_{C^1(B(p_o,\delta))} < \varepsilon$, and preserves p_o as the unique critical point in $B(p_o,\delta)$. Applying Theorem 2.2, we obtain

$$C_*(f,p_o) = H_*(W,W_-) = C_*(g,p_o).$$

The proof is completed.

This enables us to deduce to the following homotopy invariance theorem.

COROLLARY 6.1. *Let* $\{f_\sigma \in C^2(H,\mathbb{R}^1) \mid \sigma \in [0,1]\}$ *be a class of functions satisfying the P.S. condition, and possessing* p_o *as a unique critical point in a* *fixed ball* $B(p_o,\delta)$, $\delta > 0$. *Suppose that the functions* f_σ *depend on* σ *continuously in the* $C^1(B(p_o,\delta))$*-topology. Then the critical groups* $C_*(f_\sigma,p_o)$ *are independent of* σ.

Next, we shall extend this result to functions with isolated critical sets. Let us recall the definition of the Gromoll-Meyer pair of an isolated critical set, and get a variant of Lemma 6.2.

Let f,g be C^1-functions satisfying the P.S. condition. Suppose that there is an open set $\mathcal{O}$ in H, such that

$$S_f = K_f \cap \tilde{\mathcal{O}}_{df} \cap f^{-1}(\alpha,\beta), \quad S_g = K_g \cap \tilde{\mathcal{O}}_{dg} \cap g^{-1}(\alpha',\beta')$$

are isolated critical sets of f and g respectively.

LEMMA 6.2'. *Let* (W,W_-) *be a Gromoll-Meyer pair of* S_f *such that* $\mathcal{O} \subset W$. *Then there exists* $\varepsilon > 0$ *such that* (W,W_-) *is also a Gromoll-Meyer pair of* S_g *with respect to a certain pseudo-gradient vector field of* g, *provided* $\|g-f\|_{C^1(\overline{W})} < \varepsilon$.

The proof is similar to the proof of Lemma 6.2. One should only notice that both S_f and S_g are compact subsets in $\mathcal{O}$ provided by the P.S. condition. Thus we can choose an open subset $\mathcal{O}' \subset \overline{\mathcal{O}}' \subset \mathcal{O}$, such that $S_f \cup S_g \subset \mathcal{O}'$. Now the two balls $B(p_o,r/2)$ and $B(p_o,r)$ in the proof of Lemma 6.2 are replaced by $\mathcal{O}'$ and $\mathcal{O}$ respectively. The rest of the proof is the same as before.

THEOREM 6.4. *Let* $\{f_\sigma \in C^2(H,\mathbb{R}^1) \mid \sigma \in [0,1]\}$ *be a class of functions satisfying the P.S. condition. Suppose that there exist an open set* $\mathcal{O}$ *and a bounded set* N *in* H *such that* $\mathcal{O} \subset N$. *Let* K_σ *be the critical set of* f_σ, $\tilde{\mathcal{O}}_\sigma$ *the tubular set generated by* $\mathcal{O}$ *with respect to the negative gradient flow of* f_σ, $\sigma \in [0,1]$. *Assume that* $S_\sigma = K_\sigma \cap \tilde{\mathcal{O}}_\sigma \cap f_\sigma^{-1}(\alpha_\sigma,\beta_\sigma)$ *is the isolated critical set of* f_σ, *with* $S_\sigma \subset \mathcal{O}$, *and that* $\tilde{\mathcal{O}}_\sigma \cap f_\sigma^{-1}(\alpha_\sigma,\beta_\sigma) \subset N$. *If the functions* f_σ *depend on* σ *continuously in the* $C^1(N)$*-topology, then the critical groups*

$$H_*((f_\sigma)_{\beta_\sigma} \cap \tilde{\mathcal{O}}_\sigma, ((f_\sigma)_{\beta_\sigma} \backslash [S_\sigma]) \cap \tilde{\mathcal{O}}_\sigma; G)$$

are independent of σ.

PROOF. We divide $[0,1]$ into n disjoint intervals: $[0,1] = \bigcup_{i=1}^{n} I_i$, where $I_i = [a_{i-1}, a_i]$, $a_o = 0$, $a_n = 1$, such that

$$\|f_\sigma - f_{a_{i-1}}\|_{C^1(N)} < \varepsilon \qquad \forall \sigma \in I_i, \quad i = 1,\dots,n,$$

where ε is the number obtained in Lemma 6.2'.

According to the Theorems 2.6 and 1.5 we have

$$H_*((f_\sigma)_{\beta_\sigma} \cap \tilde{\mathcal{O}}_\sigma, ((f_\sigma)_{\beta_\sigma} \backslash [S_\sigma]) \cap \tilde{\mathcal{O}}_\sigma; G) \cong H_*(W_\sigma, W_{\sigma-}; G),$$

where $W_\sigma = \tilde{\mathcal{O}}_\sigma \cap f_\sigma^{-1}(\alpha'_\sigma, \beta'_\sigma)$ and $W_{\sigma-} = W_\sigma \cap f_\sigma^{-1}(\alpha'_\sigma)$, and where $\alpha'_\sigma \in (\alpha_\sigma, \inf\{f_\sigma(x) \mid x \in S_\sigma\})$, $\beta'_\sigma \in (\sup\{f_\sigma(x) \mid x \in S_\sigma\}, \beta_\sigma)$. Thus $W_\sigma \subset N$. By Lemma 6.2' and Theorem 2.6,

$$H_*(W_\sigma, W_{\sigma-}; G) \cong H_*(W_{a_{i-1}}, W_{a_{i-1}-}; G) \qquad \forall \sigma \in I_i,$$

$i = 1,2,\dots,n$. Therefore, they are all equal to $H_*(W_o, W_{o-}; G)$. The proof is finished.

REMARK 6.1. For Theorems 6.1, 6.2, refer to Marino and Prodi [53]. Corollary 6.1 first appeared in Gromoll and Meyer [39].

Theorem 6.4, i.e., the homotopy invariance of certain critical groups of isolated critical sets, is an analogue of a theorem due to Conley.

One should notice that the definition of isolated critical sets is different from that of an isolated invariant set, so that the conditions for homotopy invariance are different. The theory presented here is considerably simpler than that of Conley.

Now we turn to study the bifurcation problems.

Let H be a Hilbert space, and Ω be a neighbourhood of θ in H. Suppose that L is a bounded self-adjoint operator on H, and that $G \in C(\Omega,H)$, with $G(u) = o(\|u\|)$ at $u = \theta$. We assume that G is a potential operator, i.e., $\exists g \in C^1(\Omega,\mathbb{R}^1)$, such that $dg = G$. Find solutions of the following equation with a parameter $\lambda \in \mathbb{R}^1$:

$$Lu + G(u) = \lambda u. \tag{6.2}$$

Obviously $u = \theta$, for all $\lambda \in \mathbb{R}^1$, is a solution of (6.2). We are concerned with the nontrivial solutions of (6.2) with small $\|u\|$.

Because (6.2) is the Euler equation of a functional with parameter λ, the bifurcation phenomenon has its specific feature. We shall prove the following theorem due to Krasnoselskii [46] and Rabinowitz [67] via Morse theory, cf. [26].

THEOREM 6.5. *Suppose that* $f \in C^2(\Omega,\mathbb{R}^1)$ *with* $df(u) = Lu + G(u)$, L *being linear and* $G(u) = o(\|u\|)$ *at* $u = \theta$. *If* μ *is an isolated eigenvalue of* L *of finite multiplicity, then* (μ,θ) *is a bifurcation point for* (6.2). *Moreover, at least one of the following alternatives occurs:*

(1) (μ,θ) *is not an isolated solution of* (6.2) *in* $\{\mu\} \times \Omega$.

(2) *There is a one-sided neighbourhood* Λ *of* μ *such that for all* $\lambda \in \Lambda\backslash\{\mu\}$, (6.2) *possesses at least two distinct nontrivial solutions.*

(3) *There is a neighbourhood* I *of* μ *such that for all* $\lambda \in I\backslash\{\mu\}$, (6.2) *possesses at least one nontrivial solution.*

The proof depends upon the Lyapunov-Schmidt reduction. Let $X = \ker(L-\mu I)$, with $\dim X = n$; and let $P,P^\perp$ be the orthogonal projections onto X and $X^\perp$, respectively. Then (6.2) is equivalent to a pair of equations

$$\mu x + PG(x+x^\perp) = \lambda x \tag{6.3}$$

$$Lx^\perp + P^\perp G(x+x^\perp) = \lambda x^\perp \tag{6.4}$$

where $u = x + x^{\perp}$, $x \in X$, $x^{\perp} \in X^{\perp}$. The equation (6.4) is uniquely solvable in a small bounded neighbourhood $\mathcal{O}$ of $(\mu,\theta) \in \mathbb{R}^1 \times X$, say $x^{\perp} = \varphi(\lambda,x)$ for $(\lambda,x) \in \mathcal{O}$, where $\varphi \in C^1(\mathcal{O},X^{\perp})$. Substitute $x^{\perp} = \varphi(\lambda,x)$ into (6.3),

$$\mu x + PG(x + \varphi(\lambda,x)) = \lambda x \tag{6.5}$$

which is again a variational problem on the finite dimensional space X. Let

$$\begin{aligned} J_\lambda(x) &= f(x + \varphi(\lambda,x)) - \frac{\lambda}{2}(\|x\|^2 + \|\varphi(\lambda,x)\|^2) \\ &= \tfrac{1}{2}(\mu-\lambda)\|x\|^2 + \tfrac{1}{2}(L\varphi,\varphi) - \frac{\lambda}{2}\|\varphi\|^2 + g(x+\varphi) \end{aligned} \tag{6.6}$$

where $dg = G$, with $g(\theta) = 0$. It is easy to verify that (6.5) is the Euler equation of J_λ, and that $\varphi(\lambda,x) = o(\|x\|)$ at $x = \theta$.

The problem is reduced to find critical points of J_λ near $x = \theta$ for fixed λ near μ, where $J_\lambda \in C^1(\Omega_1,\mathbb{R}^1)$, Ω_1 is a neighbourhood of θ in X.

PROOF of Theorem 6.5. Clearly $x = \theta$ is a critical point of J_λ $\forall\lambda$ such that $(\lambda,\theta) \in \mathcal{O}$. If θ is not an isolated critical point of J_μ, which corresponds to case (1) in the theorem, then there are only two possibilities:

(i) either $x = \theta$ is a local maximum or a local minimum of J_μ;

(ii) $x = \theta$ is neither a local maximum nor a local minimum of J_μ.

<u>In case (i)</u>, suppose that θ is a local minimum of J_μ. For some $\varepsilon > 0$, $W = (J_\mu)_\varepsilon = \{x \in \Omega_1 \mid J_\mu(x) \le \varepsilon\}$ is a neighbourhood of θ, containing θ as the unique critical point. The negative gradient flow of J_μ preserves W, therefore the negative gradient flow of J_λ preserves W for $|\lambda-\mu|$ small. Since W is contractible, $\chi(M) = 1$, $x = \theta$ is a local maximum of J_λ, for $\lambda > \mu$ and J_λ is bounded from below on W; we obtain two nontrivial critical points, according to Theorem 5.1, in particular, Corollary 5.3. Therefore, for each λ in a small right hand side neighbourhood of μ, there exist at least two distinct nontrivial solutions of (6.2).

Similarly, we prove that there exist at least two distinct nontrivial solutions of (6.2) for each λ in a small left hand side neighbourhood of μ, if θ is a local maximum of J_μ.

<u>In case (ii)</u>, θ is neither a local maximum nor a local minimum of J_μ. We see that

$$C_o(J_\mu,\theta) = C_n(J_\mu,\theta) = 0, \tag{6.7}$$

according to Example 4 in Section 1. Since

$$C_o(J_\lambda,\theta) = 1, \quad \text{for } \lambda < \mu, \text{ and} \tag{6.8}$$

$$C_n(J_\lambda,\theta) = 1, \quad \text{for } \lambda > \mu; \tag{6.9}$$

we conclude that there is a neighbourhood I of μ such that for $\lambda \in I\backslash\{\mu\}$, J_λ possesses a nontrivial critical point. If not, $\exists\lambda_m \to \mu$, say $\lambda_m > \mu$, such that J_{λ_m} has the unique critical point θ, then $C_n(J_{\lambda_m},\theta) = 1$, $m = 1,2,\ldots,$ implies $C_n(J_\mu,\theta) = 1$ by Theorem 6.3. This contradicts (6.7). Similarly for $\lambda_m < \mu$. This complete the proof.

REMARK 6.2. A weaker result that (μ,θ) is a bifurcation point, was proved by a simpler argument via Theorem 6.2, cf. Berger [15].

More information on the number of distinct solutions can be obtained if we assume, in addition, that the function f is even. We confine ourselves to a special case, in which θ is an isolated local minimum (or maximum) of J_μ, and conclude that for $\lambda > \mu$ (or $\lambda < \mu$) with $|\lambda-\mu|$ small, J_λ has at least n distinct pairs of nontrivial solutions.

In fact, J_λ is even. Suppose that θ is an isolated local minimum of J_μ. For $\varepsilon > 0$, let $W = (J_\mu)_\varepsilon$; then W is invariant under the negative gradient flow of J_μ, and is contractible. Now, θ is a local maximum of J_λ for $\lambda > \mu$, with $|\lambda-\mu|$ small. There is a small ball $B(\theta,\delta) \subset \mathring{W}$, such that $-dJ_\lambda|_{\partial B(\theta,\delta)}$

directs inward in W. Let $M = W\backslash B(\theta,\delta)$, then it is a symmetric domain in $\mathbb{R}^n$ on whose boundary $-dJ_\lambda$ directs inward. Glueing the antipodal points, and deforming according to the negative gradient flow of J_μ, we have $\hat{M} \simeq P^{n-1}$, the real projective space, where $\hat{M} = M/\mathbb{Z}_2$. Thus

$$\text{cuplength of } \hat{M} = \text{cuplength of } P^{n-1} = n-1,$$

and the conclusion follows from Corollary 4.2. Similarly, we consider the case where θ is a local maximum of J_μ.

REMARK 6.3. In general, the following theorem was obtained by Fadell and Rabinowitz [35]:

THEOREM. *Let* X *be a real Hilbert space,* Ω *a neighbourhood of* θ *in* X, *and* $f \in C^2(\Omega,\mathbb{R}^1)$, *where* f *is even and* $df(u) = Lu + H(u)$ *with* L *linear and* $H(u) = o(\|u\|)$ *at* $\mu = \theta$. *Suppose that* $\mu \in \sigma(L)$ *is an isolated eigenvalue of* L *of multiplicity* $n < \infty$. *Then either* (i) (μ,θ) *is not an isolated solution of* (6.2) *in* $\{\mu\} \times X$ *or* (ii) *there exist left and right neighbourhoods* Λ_ℓ *and* Λ_r *of* μ *in* $\mathbb{R}^1$ *and integers* $k,m \geq 0$ *such that* $k+m \geq n$, *and if* $\lambda \in \Lambda_\ell$ *(resp.* Λ_μ*),* (6.2) *possesses at least* k *(resp.* m*) distinct pairs of nontrivial solutions.*

Section 7

HOMOLOGICAL CHARACTERIZATION OF THE MOUNTAIN PASS POINT

Let X be a Banach space, and let $f \in C^1(X,\mathbb{R}^1)$ be a function possessing the deformation property. The mountain pass lemma due to Ambrosetti and Rabinowitz [10] has the following variant. Let $x_o, x_1 \in X$, and let

$$F = \{\ell \in C([0,1],X) \mid \ell(0) = x_o,\ \ell(1) = x_1\}.$$

Define

$$c = \inf_{\ell \in F} \sup_{x \in \ell} f(x). \tag{7.1}$$

If $c > \max\{f(x_o), f(x_1)\}$, then c is a critical value of f (cf. K.C. Chang [23]).

One may ask if we have any information about the critical groups for critical points in K_c? First, we give a general version of the mountain pass lemma.

LEMMA 7.1. *Let* $f \in C^1(X,\mathbb{R}^1)$ *be a function possessing the deformation property. Let* c *be defined as in* (7.1) *with* $c > \max\{f(x_o), f(x_1)\}$. *Then* c *must be a critical value, and one of the following possibilities holds:* (1) c *is not isolated;* (2) K_c *is an infinite set;* (3) *there exists* $\hat{p} \in K_c$ *such that* rank $C_1(f,\hat{p}) > 0$.

PROOF. We consider the following exact sequence:

$$\cdots \to H_1(f_{c+\varepsilon}, \mathring{f}_c) \to H_o(\mathring{f}_c) \overset{i^*}{\to} H_o(f_{c+\varepsilon}) \to 0 \qquad \forall \varepsilon > 0$$

where $i: \mathring{f}_c \to f_{c+\varepsilon}$ is the injection. Since x_o, x_1 belong to a path-connected component in $f_{c+\varepsilon}$, but not in $\mathring{f}_c$, we see that

$$\ker i^* \neq 0.$$

Thus

$$H_1(f_{c+\varepsilon}, \mathring{f}_c) \neq 0.$$

It follows that $K \cap f^{-1}[c, c+\varepsilon] \neq \emptyset$, and then c is a critical value. Assume that c is an isolated critical value, and that K_c consists of a finite set, say $\{p_1, \ldots, p_n\}$. Then by (1.2),

$$H_1(f_{c+\varepsilon}, \mathring{f}_c) \cong \bigoplus_{i=1}^{n} C_1(f, p_i)$$

and the lemma is proved.

THEOREM 7.1. *Suppose that* H *is a Hilbert space, and that* $f \in C^2(H, \mathbb{R}^1)$ *satisfies the P.S. condition. Assume that* $\hat{p}$ *is a critical point of mountain pass type, i.e.,* $\hat{p}$ *is isolated and*

$$\operatorname{rank} C_1(f, \hat{p}) > 0, \tag{7.2}$$

and that either $0 \notin \sigma(d^2 f(\hat{p}))$ *or* 0 *is isolated in* $\sigma(d^2 f(\hat{p}))$, *and*

$$\dim \ker(d^2 f(\hat{p})) = 1 \quad \text{if} \quad 0 \in \sigma(d^2 f(\hat{p})). \tag{7.3}$$

Then

$$C_k(f, \hat{p}) = \begin{cases} G & \text{if} \quad k = 1 \\ 0 & \text{if} \quad k \neq 1. \end{cases}$$

PROOF. Let q be the Morse index at $\hat{p}$. If $\hat{p}$ is non-degenerate, then $C_k(f, \hat{p}) = G$ or 0 according as $k = q$ or $k \neq q$. (7.2) implies $q = 1$ and the conclusion follows. Now we assume that $\hat{p}$ is degenerate. The shifting theorem implies the equality

$$C_1(f, \hat{p}) = C_{1-q}(h, \hat{p}) \tag{7.4}$$

where $h = f|_N$ with $h(\hat{p}) = 0$, and N is the characteristic submanifold of f at $\hat{p}$. According to Lemma 7.1, $q \leqq 1$. Two cases are considered:

<u>$q = 1$</u>. By (7.2) and (7.4) we see that

$$C_0(h,\hat{p}) \neq 0. \tag{7.5}$$

Since $\hat{p}$ is an isolated critical point, and h is defined on a 1-manifold, $\hat{p}$ is a local minimum of h (see Examples 1, 2 in Section 1). Then we have

$$C_k(h,\hat{p}) = \begin{cases} G, & k = 0 \\ 0, & k \neq 0, \end{cases}$$

and hence

$$C_k(f,\hat{p}) = \begin{cases} G, & k = 1 \\ 0, & k \neq 1. \end{cases}$$

<u>$q = 0$</u>. We have

$$C_k(f,\hat{p}) = C_k(h,\hat{p}) \qquad k = 0,1,2,\ldots,$$

where h is a one-variable function. According to the Examples 1 and 3 in Section 1, the only possibility in agreement with the condition $C_1(h,\hat{p}) \neq 0$, is the case that $\hat{p}$ is a local maximum of h. Thus

$$C_k(h,\hat{p}) = \begin{cases} G, & k = 1 \\ 0, & k \neq 1. \end{cases}$$

COROLLARY 7.1. *Suppose that* $f \in C^2(H,\mathbb{R}^1)$ *satisfies the P.S. condition, and that* $df = id - T$, *where* T *is compact. If* $\hat{p}$ *is an isolated critical point of* f *which is of mountain pass type, i.e.,* $C_1(f,\hat{p}) \neq 0$, *and if further we assume that the smallest eigenvalue* λ_1 *of* $d^2f(\hat{p})$ *is simple when* $\lambda_1 = 0$. *Then we have* $\lambda_1 \leq 0$, *and*

$$\text{index}(df,\hat{p}) = -1.$$

PROOF. If $\lambda_1 > 0$, then $\hat{p}$ must be a local minimum, which implies $C_1(f,\hat{p}) = 0$.

Thus $\lambda_1 \le 0$. The rest of the corollary is a direct consequence of the theorem in view of Theorem 4.1.

REMARK 7.1. The corollary was independently obtained by Hofer [42] and G. Tian [79] but their proofs are different. In an earlier paper, A. Ambrosetti [7] proved the same conclusion under a stronger hypothesis, where $\hat{p}$ is assumed nondegenerate. The proof presented here is according to Tian [7] where K_c is assumed to consist of a single point.

REMARK 7.2. The condition (2) is essential. In fact, the monkey saddle provides a counterexample if (2) is dropped.

Finally, we turn to a variant of the mountain pass lemma, in which the P.S. condition is satisfied under a different topology. The motivation for this extension may be found in Chang [23].

THEOREM 7.2. *Suppose that* X *is a Banach space which is dense, and is continuously imbedded in a Hilbert space* H. *Assume that* $f \in C^{2-0}(H,\mathbb{R}^1)$ *is a function satisfying the P.S. condition with* $K \subset X$. *Let* $\tilde{f} = f|_X$ *possess the deformation property, let* $x_0, x_1 \in X$, *and let*

$$c = \inf_{\ell \in F} \sup_{x \in \ell} \tilde{f}(x)$$

where $F = \{\ell \in C([0,1],X) \mid \ell(0) = x_0,\ \ell(1) = x_1\}$. *If*

$$c > \max\{\tilde{f}(x_0), \tilde{f}(x_1)\},$$

then one of the following three possibilities holds: (1) c *is not isolated;* (2) K_c *is an infinite set;* (3) *there exists a point* $\hat{p} \in K_c$ *such that* rank $C_1(f,\hat{p}) > 0$.

REMARK 7.3. By the assumptions in the theorem, the critical set K of f is the same as that of $\tilde{f}$.

PROOF. Following the proof of Lemma 7.1, one obtains

(7.7) $$H_1(\tilde{f}_{c+\varepsilon}, \mathring{f}_c) \neq 0.$$

Applying Theorem 3.2, we have (see the proof of Corollary 3.2)

(7.8) $$H_1(\tilde{f}_{c+\varepsilon}, \mathring{f}_c) \cong H_1(f_{c+\varepsilon}, \mathring{f}_c).$$

Since f satisfies the P.S. condition, it follows that $K \cap f^{-1}[c, c+\varepsilon] \neq \emptyset$. The rest of the proof is the same as for Lemma 7.1.

COROLLARY 7.2. *Suppose that* f *and* $\tilde{f}$ *are defined as before, and that* $f \in C^{2-0}(H, \mathbb{R}^1)$ *satisfies the P.S. condition with* $K \subset X$. *Assume that* θ *is a strictly local minimum of* $\tilde{f}$, *and that* $\tilde{f}$ *is unbounded form below. Then* f *has at least one nontrivial critical point.*

In applications H is sometimes taken to be $H_0^1(\Omega)$ and X is the Banach space $C_0^1(\bar{\Omega})$. Corollary 7.2 is employed to connect the sub- and super-solution method with the variational methods.

Section 8

ASYMPTOTICALLY LINEAR EQUATIONS

Let H be a real Hilbert space, and let A be a bounded self-adjoint operator defined on H. According to its spectral decomposition, $H = H_+ \oplus H_o \oplus H_-$, where H_+, H_o, H_- are invariant subspaces, corresponding to the positive, zero and negative spectrum of A, respectively. Let P_+, P_o, P_- be the orthogonal projections of these subspaces.

The following assumptions are given:

(H_1) $A_\pm = A|_{H_\pm}$ has a bounded inverse on $H_\pm$;

(H_2) $\gamma \triangleq \dim(H_+ \oplus H_o) < +\infty$;

(H_3) $g \in C^2(H,\mathbb{R}^1)$ has a bounded and compact differential $dg(x)$. In addition, when $\dim H_o \neq 0$, we assume

$$g(P_o x) \to -\infty \quad \text{as} \quad \|P_o x\| \to \infty. \tag{8.1}$$

We shall study the number of critical points of the following functional:

$$f(x) = \tfrac{1}{2}(Ax,x) + g(x). \tag{8.2}$$

The following is a basic theorem.

THEOREM 8.1. *Under the assumptions* (H_1), (H_2) *and* (H_3), *if* $\{p_i\}_1^k$ *are given critical points of* f, *satisfying*

$$\text{ind } d^2f(p_i) > \gamma, \qquad i = 1,2,\ldots,k,$$

then f *has at least* k+1 *critical points.*

PROOF. 1° Firstly, we shall verify that f satisfies the P.S. condition. For $\{x_n\}_1^\infty \subset H$, $df(x_n) \to \theta$ implies that $\forall \varepsilon > 0$ $\exists N = N(\varepsilon)$ such that for $n > N$

$$|(Ax_n, x_n^\pm) + (dg(x_n), x_n^\pm)| \le \varepsilon \|x_n^\pm\|,$$

where $x_n^\pm = P_\pm x_n$. Hence $\|x_n^\pm\|$, and then (Ax_n, x_n), are bounded. Since

$$\begin{aligned}|g(P_o x_n)| &\le |g(x_n) - g(P_o x_n)| + |g(x_n)| \\ &\le m\|x_n^+ + x_n^-\| + |g(x_n)|\end{aligned}$$

where $m = \sup\{\|dg(x)\| \mid x \in H\}$. If $f(x_n)$ is bounded along this sequence, then $|g(x_n)|$, and therefore $|g(P_o x_n)|$, is bounded. Thus $\|P_o x_n\|$ is bounded. Since dg is compact, there is a subsequence x_{n_i} such that $dg(x_{n_i})$ is convergent. By

$$df(x_{n_i}) = A_+ x_{n_i}^+ + A_- x_{n_i}^- + dg(x_{n_i}) \to \theta,$$

and the boundedness of $A_\pm^{-1}$, we conclude that $x_{n_i}^\pm$ is convergent.

Since $\dim H_o$ is finite, there is a convergent subsequence $P_o x_{n_i}$. The P.S. condition is verified.

2° Denote $\varepsilon_\pm = \inf\{\|Ax_\pm\| \mid \|x_\pm\| = 1\}$, which is positive, and $R_+ = \frac{m+1}{\varepsilon_+}$. Let $M = (H_+ \cap B_{R_+}) \times (H_o \oplus H_-)$. From

$$\begin{aligned}(df(x), x_+) &= (Ax_+, x_+) - (dg(x), x_+) \\ &\ge \varepsilon_+ \|x_+\|^2 - m\|x_+\|\end{aligned}$$

we know that f has no critical point outside M, and that $-df(x)$ points inward to M on each point on ∂M.

Noticing that

$$-\tfrac{1}{2}\|A\|\|x_-\|^2 - m(\|x_-\| + R_+) + g(P_0x) \le f(x)$$
$$\le \tfrac{1}{2}\|A\|R_+^2 - \tfrac{1}{2}\varepsilon_-\|x_-\|^2 + m(\|x_-\| + R_+) + g(P_0x),$$

we obtain

$$f(x) \to -\infty \iff \|x_- + P_0x\| \to \infty \quad \text{uniformly in } x_+;$$

i.e., $\forall T > 0 \ \exists a_1 < a_2 < -T, \ R_1 > R_2 > 0$ such that

$$(H_+ \cap B_{R_+}) \times ((H_0 \oplus H_-)\backslash B_{R_1}) \subset f_{a_1} \subset (H_+ \cap B_{R_+}) \times ((H_0 \oplus H_-)\backslash B_{R_2}) \subset f_{a_2}.$$

3^0 If the conclusion of this theorem were not true, i.e., if f has only critical points $\{p_i\}_1^k$, then we could choose $T > 0$ large enough such that $\{p_i\}_1^k \not\subset f_{-T}$, i.e., $K \cap f_{-T} = \emptyset$, where K is the critical set of f. The negative gradient flow η defines a strong deformation retract $\tau_1 : f_{a_2} \to f_{a_1}$.

There is another strong deformation retract

$$\tau_2 : (H_+ \cap B_{R_+}) \times ((H_0 \oplus H_-)\backslash B_{R_2}) \to (H_+ \cap B_{R_+}) \times ((H_0 \oplus H_-)\backslash B_{R_1})$$

defined by $\tau_2 = \xi(1,.)$, where $(x_0 = P_0x)$

$$\xi(t;x_++x_0+x_-) = \begin{cases} x_+ + x_0 + x_- & \text{if } \|x_0 + x_-\| \ge R_1 \\ x_+ + \dfrac{x_0 + x_-}{\|x_0 + x_-\|}(tR_1 + (1-t)\|x_0 + x_-\|), & \|x_0 + x_-\| \le R_1. \end{cases}$$

Composing these two strong deformation retracts, i.e., $\tau = \tau_2 \circ \tau_1$, we obtain a strong deformation retract $\tau : f_{a_2} \to (H_+ \cap B_{R_+}) \oplus ((H_0 \oplus H_-)\backslash B_{R_1})$. Then we have

$$\begin{aligned} H_k(M, f_{a_2}) &\cong H_k(M, (H_+ \cap B_{R_+}) \times ((H_0 \oplus H_-)\backslash B_{R_1})) \\ &\cong H_k(H_0 \oplus H_-, (H_0 \oplus H_-)\backslash B_{R_1}) \\ &\cong H_k((H_0 \oplus H_-) \cap B_{R_1}, \partial((H_0 \oplus H_-) \cap B_{R_1})) \\ &\cong \begin{cases} G, & k = \gamma \\ 0, & k \ne \gamma. \end{cases} \end{aligned}$$

However, we have assumed

$$\text{ind } d^2f(p_i) > \gamma, \qquad i = 1,\ldots,k.$$

According to the shifting theorem, we have

$$C_k(f,p_i) = 0 \qquad \text{for } k \le \gamma.$$

Now the γth Morse inequality cannot hold, because the LHS of the Morse inequality is 0, but the RHS is 1. This is a contradiction.

Now we turn to a special case: $H_0 = \{\theta\}$. In this case the boundedness of dg in Theorem 8.1 can be dropped. Namely, we have

THEOREM 8.2. *Suppose that* $f \in C^2(H,\mathbb{R}^1)$, *and that* A *is a bounded self-adjoint operator with a bounded inverse, and a finite dimensional maximal invariant negative subspace* H_-. *Let*

$$(8.3) \qquad g(x) = f(x) - \tfrac{1}{2}(Ax,x).$$

Assume that g *has a compact differential* dg *satisfying*

$$(8.4) \qquad \|dg(x)\| = o(\|x\|) \quad \text{as} \quad \|x\| \to \infty.$$

If $\{p_i\}_1^k$ *are given critical points of* f *satisfying*

$$\text{ind } d^2f(p_i) > \dim H_-, \qquad i = 1,2,\ldots,k,$$

then f *has at least* $k+1$ *critical points.*

The following lemma is applied to reduce Theorem 8.2 to Theorem 8.1.

LEMMA 8.1. *Under the hypotheses of Theorem 8.2,* $\forall T > 0$ *there are constants* $R_2 > R_1 > T$ *and a function* $\rho \in C^\infty(\mathbb{R}^1_+)$ *such that*

$$(8.5) \qquad \rho(t) = \begin{cases} 1, & 0 \le t \le R_1 \\ 0, & t > R_2 \end{cases}$$

and

$$\tilde{f}(x) = \tfrac{1}{2}(Ax,x) + \rho(\|x\|)g(x)$$

satisfies

(8.6) $$\|d\tilde{f}(x)\| \geq 1 \quad \forall x \in B_{R_2} \backslash B_{R_1}.$$

We assume the lemma for the moment, and turn to the

PROOF of Theorem 8.2. If f has no other critical point, then we may choose $R_1 > 0$ large enough such that the critical sets of f and $\tilde{f}$ are the same. The function $\tilde{f}$ satisfies all the conditions of f in Theorem 8.1, except the compactness of $d(\rho(\|x\|)g(x))$. However, the compactness was used only in verifying the P.S. condition. Now we prove it directly for the specific function $\tilde{f}$. In fact, suppose that $d\tilde{f}(x_n) \to \theta$, then $\{x_n\}_1^\infty \subset B_{R_1}$ except for finitely many points, according to the inequality (8.6) and the invertibility of A. Since dg is compact, there is a convergent subsequence $dg(x_{n_i})$. Comparing with the assumption $df(x_n) \to \theta$, and the boundedness of A^{-1}, we obtain a convergent subsequence x_{n_i}. Thus the P.S. condition is verified; accordingly, the theorem follows from Theorem 8.1.

PROOF of Lemma 8.1. Let $\varepsilon = \frac{1}{5}\|A^{-1}\|^{-1}$. From (8.4), there is an $R_o > 0$ such that

$$\|dg(x)\| < \varepsilon\|x\| \quad \text{for } \|x\| > R_o.$$

The compactness of dg implies that $\exists M_\varepsilon > 0$ such that

$$\|dg(x)\| < \varepsilon\|x\| + M_\varepsilon \quad \forall x \in H.$$

Thus, by the mean value inequality,

$$|g(x)| < \varepsilon\|x\|^2 + M_\varepsilon\|x\| + |g(\theta)|.$$

Let

$$R_1 > \max\{T, R_o, \tfrac{1}{\varepsilon}(4M_\varepsilon + 3)\}, \quad \lambda = \max\{1, |g(\theta)|\}R_1$$

and let $\varphi \in C^\infty(\mathbb{R}^1_+)$ be a function satisfying $0 \leq \varphi(t) \leq 1$ and

$$\varphi(t) = \begin{cases} 1 & \text{for } t \leq 0 \\ 0 & \text{for } t \geq 1 \end{cases}$$

as well as

$$\max|\varphi'(t)| \le \frac{3}{2} .$$

Let us define

$$\rho(t) = \varphi\left(\frac{t-R_1}{\lambda}\right)$$

and $R_2 = R_1 + \lambda$, then (8.5) is satisfied, and then

$$\begin{aligned}\|d\tilde{f}(x)\| &= \|Ax + \frac{1}{\lambda}\varphi'\left(\frac{\|x\|-R_1}{\lambda}\right)g(x)\frac{x}{\|x\|} + \varphi\left(\frac{\|x\|-R_1}{\lambda}\right)dg(x)\| \\ &\ge \|A^{-1}\|^{-1}\|x\| - (\varepsilon\|x\| + M_\varepsilon) - \frac{3}{2\lambda}(\varepsilon\|x\|^2 + M_\varepsilon\|x\| + |g(\theta)|) \\ &\ge (5\varepsilon - \varepsilon - \frac{3\varepsilon}{2\lambda}\|x\|)\|x\| - (M_\varepsilon + \frac{3}{2\lambda}\|x\|M_\varepsilon + \frac{3}{2\lambda}|g(\theta)|) \\ &\ge \varepsilon\|x\| - (4M_\varepsilon + 2) \ge 1\end{aligned}$$

$\forall x \in B_{R_2}\backslash B_{R_1}$, whence $\frac{1}{\lambda}\|x\| \le 2$.

COROLLARY 8.1. *Suppose that* $f \in C^2(\mathbb{R}^n,\mathbb{R}^1)$ *and that* A_∞ *is an* $n \times n$ *non-singular symmetric matrix such that* $g(x) = f(x) - \frac{1}{2}(A_\infty x,x)$ *satisfies*

$$\|dg(x)\| = o(\|x\|) \quad \textit{as} \quad \|x\| \to \infty.$$

If θ *is a critical point of* f *with the condition*

$$\operatorname{ind}(A_\infty) \notin [m_-,m_0+m_-],$$

where $m_- = \operatorname{ind} d^2f(\theta)$ *and* $m_0 = \dim\ker d^2f(\theta)$, *then* f *has at least one non-trivial critical point.*

PROOF. If $\operatorname{ind}(A_\infty) < m_-$, then the conclusion is a direct consequence of Theorem 8.2. If $\operatorname{ind}(A_\infty) > m_- + m_0$, then

$$\begin{aligned}\operatorname{ind}(-A_\infty) &= n - \operatorname{ind}(A_\infty) \\ &< n - (m_-+m_0) = \operatorname{ind}(d^2(-f)(\theta)).\end{aligned}$$

Applying Theorem 8.2 to the function $-f$, we obtain the conclusion.

REMARK 8.1. Corollary 8.1 was first proved by Amann and Zehnder [6], where the Conley index, a kind of generalized Morse index, was applied. Later on, Chang [21] extended it to Theorem 8.2 with $H = \mathbb{R}^n$, by the Morse theory under general boundary condition, cf. Morse and Cairns [57]. The special case of Theorem 8.1, with $H = \mathbb{R}^n$, was given by Liu [51], who applied a homotopy argument on the one point compactification space $\mathbb{R}^n \cup \{\infty\}$. The proof presented here is due to Chang [25].

COROLLARY 8.2. *Suppose that* $f \in C^2(H,\mathbb{R}^1)$, *and that* A *is a bounded self-adjoint operator with a bounded inverse, and a finite dimensional maximal invariant negative subspace* H_-. *Let*

$$g(x) = f(x) - \tfrac{1}{2}(Ax,x).$$

Assume that g *has a compact differential* dg *satisfying*

$$\|dg(x)\| = o(\|x\|) \quad \text{as} \quad \|x\| \to \infty.$$

(i) *If* f *is nondegenerate, then* f *has an odd number of critical points;*

(ii) *If two distinct nondegenerate critical points of* f *are given, then there exist at least three distinct critical points of* f.

PROOF. Basically, the proof is the same as that of Theorem 8.2. First we choose a suitable $R_1 > 0$ such that the corresponding function $\tilde{f}$ defined in Lemma 8.1 possesses the same critical set as the function f. Second, we can choose a real number a, such that $\tilde{f}$ has no critical point in the level set f_a; the proof of this fact is the same as in section two of Theorem 8.1. Lastly, for $M = (H_+ \cap B_{R_1}) \times H_-$, we have

$$H_k(M,f_a) \cong \begin{cases} G, & k = \gamma \\ 0, & k \neq \gamma \end{cases}$$

where $\gamma = \dim H_-$. The two conclusions follow from the Morse identity (1.7).

REMARK 8.2. Corollary 8.2 is taken from Chang [21].

Next, we turn to a variant of Theorem 8.1 which provides more information on the number of critical points, if the functional f is defined on H attached by a compact manifold V.

THEOREM 8.3. *Let* A *be a bounded self-adjoint operator defined on* H, *satisfying the assumptions* (H_1) *and* (H_2). *Let* V^n *be a* C^2*-compact manifold without boundary, and let* $g \in C^2(H \times V^n, \mathbb{R}^1)$ *be a function having a bounded and compact* dg, *satisfying*

$$g(P_o x, v) \to -\infty \quad \text{as} \quad \|P_o x\| \to +\infty \quad \text{if} \quad \dim H_o \neq 0.$$

Then the function

$$f(x,v) = \tfrac{1}{2}(Ax,x) + g(x,v)$$

has at least $\text{cuplength}(V^n) + 1$ *critical points.*

If further, f *is nondegenerate, then* f *has at least* $\sum_{i=0}^{n} \beta_i(V^n)$ *critical points, where* $\beta_i(V^n)$ *is the* i^{th} *Betti number of* V^n, $i = 0,1,\ldots,n$.

PROOF. The P.S. condition of f is verified in the same way as in Theorem 8.1.

Similarly, there is $R_+ > 0$ such that f has no critical point outside $M \triangleq (H_+ \cap B_{R_+}) \times (H_o \oplus H_-) \times V^n$, and that

$$f(x,v) \to -\infty \iff \|x_o + x_-\| \to \infty \quad \text{uniformly in } M.$$

Suppose that the number of critical points of f is less than $\ell+1$, where $\ell = \text{cuplength}(V^n)$. Then we may choose a real number a such that the critical set of f does not intersect with f_a. Hence there is an $R > 0$ such that

$$\begin{aligned} H_*(M,f_a) &\cong H_*(M,(H_+ \cap B_{R_+}) \times ((H_o \oplus H_-)\backslash B_R) \times V^n) \\ &\cong H_*((H_o \oplus H_-) \cap B_R, \partial((H_o \oplus H_-) \cap B_R)) \otimes H_*(V^n) \end{aligned}$$

by the Künneth formula. Thus

$$H_k(M,f_a) \simeq H_{k-\gamma}(V^n), \qquad k = 0,1,2,\ldots \tag{8.7}$$

It is easily seen that

$$H^*(M) \simeq H^*(V^n). \tag{8.8}$$

There exist $\omega_1,\ldots,\omega_\ell \in H^*(M)$ such that $\omega_1 \cup \ldots \cup \omega_\ell \neq 0$. According to the isomorphisms (8.7) and (8.8), we have $[z_1] \in H_*(M,f_a)$ such that the duality

$$[[z_1],\omega_1 \cup \ldots \cup \omega_\ell] \neq 0.$$

Since the cap product $\cap : H_*(M,f_a) \times H^*(M) \to H_*(M,f_a)$ is the dual of the cup product, we have $\ell+1$ nontrivial singular chains in $H_*(M,f_a)$ defined by

$$[z_{j+1}] = [z_j] \cap \omega_j, \qquad j = 1,2,\ldots,\ell.$$

Applying Theorem 4.3, there are at least $\ell+1$ distinct critical points. This contradiction proves the first assertion.

If f is a Morse function, by the same argument we may restrict ourselves to the case $K \cap f_a = \emptyset$, then f has at least

$$\sum_{k=0}^{\infty} \operatorname{rank} H_k(M,f_a) = \sum_{j=0}^{\infty} \operatorname{rank} H_j(V^n) = \sum_{j=0}^{n} \beta_j(V^n)$$

critical points, provided by the Morse inequalities.

REMARK 8.3. Theorem 8.3 is taken from Chang [25].

Now we turn to applications to some problems in analysis, where certain unbounded operators are encountered. We need a finite dimensional reduction.

Let H be a real Hilbert space, and let A be a self-adjoint operator with domain $D(A) \subset H$. Let F be a potential operator with $\Phi \in C^1(H,\mathbb{R}^1)$, $F = d\Phi$, $\Phi(\theta) = 0$. Assume that

(A) There exist real numbers $\alpha < \beta$ such that $\alpha,\beta \notin \sigma(A)$, and that $\sigma(A) \cap [\alpha,\beta]$ consists of at most finitly many eigenvalues of finite multiplicities.

(F) F is Gateaux differentiable in H, which satisfies

$$\|df(u) - \frac{\alpha+\beta}{2} I\| \le \frac{\beta-\alpha}{2} \qquad \forall u \in H.$$

The problem is to find the solutions of the following equation:

$$Ax = F(x) \qquad x \in D(A). \tag{8.9}$$

A Lyapounov-Schmidt procedure is applied for a finite dimensional reduction.

Let

$$P_o = \int_\alpha^\beta dE_\lambda, \quad P_+ = \int_\beta^{+\infty} dE_\lambda, \quad P_- = \int_{-\infty}^{\alpha} dE_\lambda,$$

where $\{E_\lambda\}$ is the spectral resolution of A, and let

$$H_o = P_oH, \quad H_\pm = P_\pm H.$$

Without loss of generality, we may assume that $\alpha = -\beta$, and $0 \notin \sigma(A)$. We assume further the following condition:

(D) $\Phi \in C^2(V,\mathbb{R}^1)$, where $V = D(|A|^{\frac{1}{2}})$, with the graph norm:

$$\|x\|_V = \||A|^{\frac{1}{2}}x\|_H.$$

We decompose the space V as follows:

$$V = V_o \oplus V_- \oplus V_+,$$

where

$$V_o = |A|^{-\frac{1}{2}}H_o, \quad V_\pm = |A|^{-\frac{1}{2}}H_\pm.$$

For each $u \in H$, we have the decomposition:

$$u = u_+ + u_o + u_-$$

where $u_o \in H_o$, $u_\pm \in H_\pm$, and let $x = x_+ + x_o + x_- \in V$, where

$$x_o = |A|^{-\frac{1}{2}}u_o, \quad x_\pm = |A|^{-\frac{1}{2}}u_\pm.$$

Thus we have

$$\|x_\pm\|_{V_\pm} = \|u_\pm\|_{H_\pm}, \quad \|x_o\|_{V_o} = \|u_o\|_{H_o}$$

and that $V_\pm$, V_o are isomorphic to $H_\pm$ and H_o, respectively.

Now we define a functional on H as follows:

(8.10) $$f(u) = \tfrac{1}{2}(\|u_+\|^2 + \|E_+u_o\|^2 - \|E_-u_o\|^2 - \|u_-\|^2) - \Phi(x),$$

where $E_+ = \int_0^\infty dE_\lambda$, and $E_- = \int_{-\infty}^0 dE_\lambda$.

The Euler equation of this functional is the system:

$$\begin{cases} u_\pm = \pm|A|^{-\frac{1}{2}}P_\pm F(x) & (8.11)\\ E_\pm u_o = \pm|A|^{-\frac{1}{2}}E_\pm P_o F(x). & (8.12)\end{cases}$$

Thus $x = x_+ + x_o + x_-$ is a solution of (8.9) if and only if $u = u_+ + u_o + u_-$ is a critical point of f. However, the system (8.11) is reduced to

$$Ax_\pm = P_\pm F(x_+ + x_- + x_o)$$

which is equivalent to

(8.13) $$x_\pm = A^{-1}P_\pm F(x_+ + x_- + x_o).$$

By assumption (D), $F \in C^1(V,V)$, and by assumption (F)

$$\|F(u) - F(v)\|_H \leq \beta\|u-v\|_H \qquad \forall u,v \in H.$$

Furthermore, there is a $\gamma > \beta$ such that

(8.14) $$\|A^{-1}|_{H_+\oplus H_-}\| \leq \frac{1}{\gamma}$$

by assumption (A). We shall prove that the operator $\mathcal{F} = A^{-1}(P_+ + P_-)F \in C^1(V,V)$ is contractible with respect to variables in $V_+ \oplus V_-$. In fact, $\forall x = x_+ + x_- + z$, $y = y_+ + y_- + z$, for fixed $z \in V_o$,

$$\begin{aligned}\|\mathcal{F}(x) - \mathcal{F}(y)\|_V &= \||A|^{-\frac{1}{2}}(P_+ + P_-)(F(x) - F(y))\|_H\\ &\leq \||A|^{-\frac{1}{2}}(P_+ + P_-)\|_{\mathcal{B}(H)}\|F(x) - F(y)\|_H\\ &\leq \beta\||A|^{-\frac{1}{2}}(P_+ + P_-)\|\|(x_+ + x_-) - (y_+ + y_-)\|_H\end{aligned}$$

$$\leq \beta/\sqrt{\gamma}\| |A|^{-\frac{1}{2}}(P_{+}+P_{-})\| \|x-y\|_V$$

because $\|x_{\pm}\|_H = \| |A|^{-\frac{1}{2}}u \|_H \leq \frac{1}{\sqrt{\gamma}} \|u_{\pm}\|_H = \frac{1}{\sqrt{\gamma}} \|x_{\pm}\|_V$. But from (8.4),

$$\| |A|^{-\frac{1}{2}}(P_{+}+P_{-})\| \leq \frac{1}{\sqrt{\gamma}} .$$

We obtain

$$\|F(x) - F(y)\|_V \leq \frac{\beta}{\gamma} \|x-y\|_V .$$

The implicit function theorem can be applied, yielding a solution $x_{\pm}(x_0)$, for fixed $x_0 \in V_0$, such that $x_{\pm} \in C^1(V_0,V_{\pm})$. Since $\dim V_0$ is finite, all topologies on V_0 are equivalent. We have $u_{\pm}(x_0) = |A|^{\frac{1}{2}}x_{\pm}(x_0) \in C^1(H_0,H\ )$, which solves the system (8.11).

Let

$$a(x_0) = f(u_{+}(x_0) + u_{-}(x_0) + u_0(x_0))$$

where $u_0(x_0) = |A|^{\frac{1}{2}}x_0$ and let $z = x_0$. We have

$$\begin{aligned} a(z) &= \tfrac{1}{2}(\|A^{\frac{1}{2}}x_{+}\|^2 + \|A^{\frac{1}{2}}E_{+}z\|^2 - \|(-A)^{\frac{1}{2}}x_{-}\|^2 - \|(-A)^{\frac{1}{2}}E_{-}z\|^2) - \Phi(x) \\ &= \tfrac{1}{2}(Ax(z),x(z)) - \Phi(x(z)) \end{aligned}$$

where $x(z) = \xi(z) + z$, $\xi(z) = x_{+}(z) + x_{-}(z) \in D(A)$. Noticing that

$$d\xi(z) = A^{-1}(P_{+}+P_{-})F'(x(z))dx(z)$$

by (8.13), one sees $d\xi(z) \in D(A)$, and that

$$Ad\xi(z) = (I - P_0)F'(x(z))dx(z).$$

Thus

(8.15) $da(z) = (dx(z))^*[Ax(z) - F(x(z))] = Az - P_0F(x(z)) = Ax(z) - F(x(z))$

and

(8.16)
$$\begin{aligned} d^2a(z) &= [A - F'(x(z))]dx(z) \\ &= A|_{H_0} - P_0F'(x(z))dx(z). \end{aligned}$$

In summary, we proved

THEOREM 8.4. *Under the assumptions* (A), (F) *and* (D), *there is a one-one correspondence:*

$$z \mapsto x = x(z) = x_+(z) + x_-(z) + z$$

between the critical points of the C^2-*function* $a \in C^2(H_o, \mathbb{R}^1)$ *with the solutions of the operator equation:*

$$Ax = F(x) \qquad x \in D(A).$$

Now we turn to the asymptotic behaviour of the function a.

LEMMA 8.2. *Under the assumptions* (A), (F) *and* (D), *we assume further that there is a bounded self-adjoint operator* F_∞ *satisfying*

$$(F_\infty)\begin{cases} \text{(i)} & P_oF_\infty = F_\infty P_o; \\ \text{(ii)} & \|F(u) - F_\infty u\| = o(\|u\|) \quad \text{as} \quad \|u\| \to \infty; \\ \text{(iii)} & 0 \notin \sigma(A-F_\infty). \end{cases}$$

Then we have that

(1) $\xi(z) = o(\|z\|)$ *as* $\|z\| \to \infty$;

(2) *the function* $a(z)$ *is asymptotically quadratic with asymptotics* $A - F_\infty|_{H_o}$, *i.e.*,

$$\|da(z) - (A-F_\infty)z\| = o(\|z\|) \quad \text{as} \quad \|z\| \to \infty.$$

PROOF. By (8.13), we obtain

$$A\xi(z) = (I-P_o)F(x(z)). \tag{8.17}$$

Since P_o commutes with F_∞, we have

$$(A-F_\infty)\xi(z) = (I-P_o)[F(x(z)) - F_\infty x(z)].$$

Hence, $\forall\varepsilon > 0$ there exists $R > 0$ such that

$$\begin{aligned} \|\xi(z)\| &\leq \|(A-F_\infty)^{-1}\|\|F(x(z)) - F_\infty\, x(z)\| \\ &< \varepsilon C(\|z\| + \|\xi(z)\|), \quad \text{if} \quad \|z\| > R \end{aligned}$$

where $C = \|(A-F_\infty)^{-1}\|$; it follows that

$$\|\xi(z)\| = o(\|z\|).$$

By (8.15) we have

$$\begin{aligned}\|da(z) - (A-F_\infty)z\| &= \|Az - P_0F(x(z)) - (A-F_\infty)z\| \\ &\leqq \|F(x(z)) - F_\infty x(z)\| + \|F_\infty x(z) - F_\infty z\| \\ &= o(\|x(z)\|) = o(\|z\|) \quad \text{as} \quad \|z\| \to \infty.\end{aligned}$$

LEMMA 8.3. *Under the assumptions* (A), (F) *and* (D), *we assume that* $F(\theta) = \theta$.

(1) *If there is a self-adjoint operator* $C_0^- \in L(H,H)$ *which commutes with* P_0 *and* P_-, *such that*

$$\min(\sigma(A) \cap [\alpha,\beta])I \leq C_0^- \leq F'(\theta),$$

then

$$a(z) \leq \tfrac{1}{2}((A-C_0^-)z,z) + o(\|z\|^2) \quad \textit{as} \quad \|z\| \to 0.$$

(2) *If there is a self-adjoint operator* $C_0^+ \in L(H,H)$ *which commutes with* P_0 *and* P_+, *such that*

$$F'(\theta) \leq C_0^+ \leq \max(\sigma(A) \cap [\alpha,\beta])I,$$

then

$$a(z) \geq \tfrac{1}{2}((A-C_0^+)z,z) + o(\|z\|^2) \quad \textit{as} \quad \|z\| \to 0.$$

PROOF. By definition and (8.15),

$$\begin{aligned}a(z) &= \tfrac{1}{2}(Ax(z),x(z)) - \Phi(x(z)) \\ &= \tfrac{1}{2}(Aq,q) - \Phi(q) + \tfrac{1}{2}(Ax_+(z),x_+(z)) - (\Phi(x(z)) - \Phi(q)),\end{aligned}$$

where $q = x_-(z) + z$. We shall prove that

$$\tfrac{1}{2}(Ax_+(z),x_+(z)) - (\Phi(x(z)) - \Phi(q)) \leq 0,$$

that is,

$$a(z) \leq \tfrac{1}{2}(Aq,q) - \Phi(q). \tag{8.18}$$

In fact,

$$\Phi(x(z)) - \Phi(q) - \tfrac{1}{2}(Ax_+(z),x_+(z))$$

$$= \tfrac{1}{2}(Ax_+(z),x_+(z)) + \int_0^1 (F(tx_+(z)+q) - F(x_+(z)+q),x_+(z))\,dt$$

$$\geq \frac{\beta}{2}\|x_+(z)\|^2 - \int_0^1 \beta(1-t)\,dt\|x_+(z)\|^2 = 0.$$

However,

$$|\Phi(q) - \tfrac{1}{2}(F'(\theta)q,q)| = \left|\int_0^1 (F(tq) - F'(\theta)tq,q)\,dt\right|$$

$$\leq \tfrac{1}{2} \sup_{0<t<1} \|F'(tq) - F'(\theta)\|_{L(V,V)}\|q\|_V^2,$$

that is,

$$-\Phi(q) \leq -\tfrac{1}{2}(F'(\theta)q,q) + o(\|q\|_V^2). \tag{8.19}$$

But, $x_- \in C^1(V_0,V_-)$; if $\|z\| \to 0$, then $\|x_-(z)\|_{V_-} = O(\|z\|_{V_0})$ because $x_-(\theta) = \theta$. Thus

$$\|q\|_V = O(\|z\|). \tag{8.20}$$

Substituting (8.19) and (8.20) into (8.18), we obtain

$$a(z) \leq \tfrac{1}{2}((A-F'(\theta))q,q) + o(\|z\|^2)$$

$$\leq \tfrac{1}{2}((A-C_0^-)q,q) + o(\|z\|^2)$$

$$= \tfrac{1}{2}((A-C_0^-)x_-(z),x_-(z)) + \tfrac{1}{2}((A-C_0^-)z,z) + o(\|z\|^2)$$

as $\|z\| \to 0$. Let

$$\alpha_- = \min\{\sigma(A) \cap [\alpha,\beta]\}.$$

Then by assumption,

$$\alpha_- I \leq C_0^-.$$

This implies

$$((A-C_0^-)x_-(z),x_-(z)) \leq ((A-\alpha_- I)x_-(z),x_-(z)) \leq 0;$$

therefore

$$a(z) \le \tfrac{1}{2}((A-C_o^-)z,z) + o(\|z\|^2) \quad \text{as} \quad z \to \theta.$$

Similarly, we prove the second assertion.

Finally, we apply Corollary 8.1 to solve the operator equation (8.9) under the conditions (A), (F), (D) and (F_∞).

For a symmetric matrix B, let $m^\pm(B)$ be the dimension of the maximal positive/negative subspace.

THEOREM 8.5. *Under the assumptions* (A), (F), (D) *and* (F_∞), *we assume* $F(\theta) = \theta$. *If one of the following conditions holds:*

(1) *there exists a bounded self-adjoint* C_o^-, *commuting with* P_o *and* P_-, *such that*

$$\min\{\sigma(A) \cap [\alpha,\beta]\}I \le C_o^- \le F'(\theta)$$

and

$$m^-(A-C_o^-|_{H_o}) > m^-(A-F_\infty|_{H_o});$$

(2) *there exists a bounded self-adjoint* C_o^+, *commuting with* P_o *and* P_+, *such that*

$$F'(\theta) \le C_o^+ \le \max\{\sigma(A) \cap [\alpha,\beta]\}I$$

and

$$m^+(A-C_o^+|_{H_o}) > m^+(A-F_\infty|_{H_o});$$

then there exists at least one nontrivial solution of the equation (8.9).

PROOF. By Theorem 8.4, the problem (8.9) is reduced to finding critical points of the function $a \in C^2(H_o,\mathbb{R}^1)$. According to Lemma 8.2, a is an asymptotically quadratic function with a nonsingular symmetric matrix $A - F_\infty|_{H_o}$ as asymptotics. By Lemme 8.3, condition (1) means that $d^2a(\theta)$ is negative on the subspace Z_- on which $A - C_o^-$ is negative. Thus

$$m^-(d^2a(\theta)) \ge m^-(A-C_o^-|_{H_o}) > m^-(A-F_\infty|_{H_o}).$$

Similarly, condition (2) means that

$$m^+(d^2a(\theta)) \geq m^+(A-C_o^+|_{H_o}) > m^+(A-F_\infty|_{H_o}).$$

In this case,

$$\begin{aligned} m^-(A-F_\infty|_{H_o}) &= \dim H_o - m^+(A-F_\infty|_{H_o}) \\ &> \dim H_o - m^+(d^2a(\theta)) \\ &= m^-(d^2a(\theta)) + \dim \ker(d^2a(\theta)). \end{aligned}$$

Both cases imply that

$$m^-(A-F_\infty|_{H_o}) \notin [m^-(d^2a(\theta)), m^-(d^2a(\theta)) + \dim \ker(d^2a(\theta))].$$

The conclusion follows from Corollary 8.1.

REMARK 8.4. The finite dimensional reduction method presented here is a modification of a method due to Amann [2], and Amann and Zehnder [5], which was based on the so-called saddle point reduction method. In this lecture, avoiding the use of monotone operators and a dull verification of the implicit function theorem, we change a little of the assumptions, and thereby gain a considerable simplification of the reduction theory.

Section 9

MULTIPLE SOLUTIONS OF ELLIPTIC BOUNDARY VALUE PROBLEMS

In this section, several typical problems of semilinear elliptic equations are studied. We shall apply the critical point theorems obtained in previous sections to estimate the number of solutions of these equations.

Let Ω be a bounded open domain in $\mathbb{R}^n$ with smooth boundary $\partial\Omega$. The following BVP (boundary value problem) is studied:

$$(9.1)\qquad \begin{cases} -\Delta u = g(x,u) & x \in \Omega \\ u|_{\partial\Omega} = 0 \end{cases}$$

where g is a given function. Depending on the behaviour of g, we shall ask: How many solutions u are there?

9.1. A resonance problem

Assume that

$$(9.2)\qquad g(x,\xi) = \hat{\lambda}\xi + \phi(x,\xi)$$

where $\hat{\lambda} \in \sigma(L)$, L is the Laplacian $-\Delta$ with Dirichlet boundary condition, and $\phi \in C^1(\bar{\Omega}\times\mathbb{R}^1,\mathbb{R}^1)$ satisfying

(1) $\phi(x,\xi)$ is bounded, i.e., there exists a constant m such that $|\phi(x,\xi)| \le m$;

(2) the Landesman-Lazer condition holds, i.e.,

$$\int_\Omega \Phi(x, \sum_{i=1}^N t_i\psi_i(x))\, dx \to -\infty \quad \text{as} \quad \sum_{i=1}^N |t_i|^2 \to \infty$$

where

$$\Phi(x,\xi) = \int_0^\xi \phi(x,s)\, ds$$

and $\operatorname{span}\{\psi_1,\ldots,\psi_N\} = \ker(L - \hat\lambda I)$.

THEOREM 9.1. *Let* g *be as in* (9.2). *In addition, we assume that*

(9.3) $$|\phi'_\xi(x,\xi)| \le C_1 + C_2|\xi|^{\frac{4}{n-2}}$$

where C_1 *and* C_2 *are constants. Then the* BVP (9.1) *has a solution. If further, we assume that*

$$\phi(x,0) = 0, \quad \phi'_\xi(x,0) > \lambda_+ - \hat\lambda \qquad \forall x \in \bar\Omega$$

where $\hat\lambda < \lambda_+$ *are consecutive eigenvalues in* $\sigma(L)$, *then the* BVP (9.1) *has a nontrivial solution.*

PROOF. Define a functional on $H^1_0(\Omega)$,

(9.4) $$J(u) = \int_\Omega \{\tfrac{1}{2}(|\nabla u|^2 - \hat\lambda u^2) - \Phi(x,u)\}\, dx,$$

then (9.1) is the Euler equation of $J \in C^2(H,\mathbb{R}^1)$. Let

$$A = id - \hat\lambda L^{-1}$$

and let

$$h(u) = -\int_\Omega \Phi(x,u)\, dx,$$

the assumptions (H_1), (H_2) and (H_3) of Theorem 8.1 are all satisfied.

The first conclusion follows directly from that theorem with $k = 0$.

Now we turn to the second one. Noticing that

$$d^2J(\theta) = id - \hat\lambda L^{-1} - L^{-1}\phi'_\xi(x,0)$$

we shall prove that

$$d^2J(\theta) \le id - \lambda_+ L^{-1}.$$

In fact, $\forall v \neq \theta$

$$(\{id - \lambda_+ L^{-1} - d^2J(\theta)\}v,v) = \int_\Omega (\hat{\lambda} + \phi_\xi'(x,0) - \lambda_+)v^2 > 0,$$

where $(\ ,\ )$ is the inner product on $H_0^1(\Omega)$. And hence

$$\text{ind } d^2J(\theta) \ge \dim \bigoplus_{\lambda \le \lambda_+} \ker(id - \lambda L^{-1}) > \dim \bigoplus_{\lambda \le \hat{\lambda}} \ker(id - \lambda L^{-1})$$

$$= \text{ind}(A) + \dim \ker(A) \overset{\Delta}{=} \gamma.$$

By Theorem 8.1 the proof is completed.

REMARK 9.1. Actually, the condition (9.3) can be dropped. For the first conclusion, it is easily seen that the condition $\phi \in C^1$ is not needed either. Because in Theorem 8.1, if $k = 0$, we only want to assume $g \in C^1(H,\mathbb{R}^1)$. As to the second conclusion, let

$$E_m = \bigoplus_{\lambda \le \lambda_m} \ker(id - \lambda L^{-1}), \qquad m = 1,2,\ldots$$

and let $E = \bigcup_{m=1}^{\infty} E_m$. Then $E_1 \subset E_2 \subset \ldots$, and E is dense in $H_0^1(\Omega)$. The topology of E is defined by the inductive limit of those of E_m, i.e.,

$$E = \varinjlim E_m.$$

For each pair of open subsets (A,B) in E with $B \subset A$, it is known that

$$H_*(A,B;G) = \varinjlim H_*(A_m,B_m;G)$$

where $A_m = A \cap E_m$, and $B_m = B \cap E_m$. According to Theorem 3.2,

$$C_*(J,\theta) \cong C_*(\tilde{J},\theta) \cong \varinjlim C_*(\tilde{J}|_{E_m},\theta)$$

where $\tilde{J} = J|_E$. Since J is C^2 on the space $C_0^1(\bar{\Omega})$ (but not on $H_0^1(\Omega)$), so that $J|_{E_m}$ is C^2 and the Morse index of the Hessian of $J|_{E_m}$ at θ is not less than γ for large m, this implies that

$$C_k(\tilde{J},\theta) = 0, \qquad k \le \gamma,$$

so that

$$C_k(J,\theta) = 0, \qquad k \le \gamma .$$

Now Theorem 8.1 holds, and then it can be applied to our problem.

9.2. Nonresonance problems

THEOREM 9.2 (Amann-Zehnder). *Suppose that there exists* $\lambda_\infty \notin \sigma(L)$ *such that*

(9.5) $$\lim_{|\xi|\to\infty} \frac{g(x,\xi)}{\xi} = \lambda_\infty \text{ uniformly in } x \in \bar{\Omega} .$$

Assume that

$$g(x,0) = 0$$

and that there exist $\alpha < \beta$ *such that*

(9.6) $$\alpha \le g'_\xi(x,\xi) \le \beta.$$

If one of the following conditions holds:

$$g'_\xi(x,0) \le \bar{\lambda} - \varepsilon < \bar{\lambda} < \lambda_\infty \text{ or } g'_\xi(x,0) \ge \bar{\lambda} + \varepsilon > \bar{\lambda} > \lambda_\infty,$$

where $\bar{\lambda} \in \sigma(A)$, *then the* BVP (9.1) *has at least one nontrivial solution.*

PROOF. On the Hilbert space $H = L^2(\Omega)$, let $A = -\Delta$ with domain $D(A) = H^2 \cap H^1_0(\Omega)$, and let

$$\Phi(u) = \int_\Omega G(x,u(x))\, dx,$$

where

$$G(x,\xi) = \int_0^\xi g(x,t)\, dt.$$

Without loss of generality we may assume that α, β and $\gamma = \frac{1}{2}(\alpha+\beta) \notin \sigma(A)$. Thus $F = d\Phi$ is Gateaux differentiable, with

$$\|dF - \gamma I\| \le \tfrac{1}{2}(\beta-\alpha)$$

by (9.6). The conditions (A) and (F) in Section 8 are satisfied. Thus the BVP (9.1) is reduced to problem (8.9).

Now we turn to the condition (D) in Section 8. By the interpolation theorem (cf. Lions and Magenes [50]):

$$D(|A - \gamma I|^{\frac{1}{2}}) = D(A^{\frac{1}{2}}) \hookrightarrow H^1(\Omega) \tag{9.7}$$

and the Hölder inequality:

$$\left|\int_\Omega g'(x,u)\cdot v\cdot w\, dx\right| \leq \left(\int_\Omega |g'(x,u)|^\gamma\, dx\right)^{\frac{1}{\gamma}} \left(\int_\Omega |v|^p\, dx\right)^{\frac{1}{p}} \left(\int_\Omega |w|^p\, dx\right)^{\frac{1}{p}} \tag{9.8}$$

where

$$\frac{1}{\gamma} + \frac{2}{p} = 1, \quad p = \frac{2n}{n-2}$$

we obtain $\Phi \in C^2(D(|A - \gamma I|^{\frac{1}{2}}), \mathbb{R}^1)$, i.e., (D) is satisfied.

Let $F_\infty = \lambda_\infty I$; then by (9.5), $\forall \varepsilon > 0$ there exists a constant $C_\varepsilon > 0$ such that

$$|g(x,\xi) - \lambda_\infty \xi| < \varepsilon|\xi| + C_\varepsilon$$

that is,

$$\|F(u) - F_\infty u\|_{L^2} \leq \varepsilon\|u\|_{L^2} + C_\varepsilon\ \mathrm{mes}(\Omega)^{\frac{1}{2}}.$$

Therefore

$$\|F(u) - F_\infty u\| = o(\|u\|) \quad \text{as} \quad \|u\| \to \infty.$$

The condition (F_∞) in Section 8 is also satisfied. Finally, let $C_0^+ = \overline{\lambda} I$ or $C_0^- = \overline{\lambda} I$, then all conditions in Theorem 8.5 are satisfied, from which we proved our theorem.

REMARK 9.2. The same argument can be applied to study the asymptotically linear wave equations:

$$\begin{cases} (\partial_t^2 - \partial_x^2)u = g(x,t;u(x,t)) & (x,t) \in (0,\pi) \times [0,2\pi] \\ u(0,t) = u(\pi,t) = 0 \\ u(x,t) \text{ is } 2\pi\text{-periodic in } t \end{cases} \tag{9.1'}$$

where g satisfies the same conditions as in Theorem 9.2, with either $\beta > \alpha > 0$ or $\alpha < \beta < 0$.

The main distinction in the proof lies in that the imbedding

$$D(|A - \gamma|^{\frac{1}{2}}) \hookrightarrow H^{\frac{1}{2}}((0,\pi) \times [0,2\pi])$$

is replacing the imbedding (9.7).

In fact, by Fourier series expansion, let

$$u(x,t) = \sum_{j=-\infty}^{\infty} \sum_{k=1}^{\infty} c_{jk} \sin kx e^{ijt}, \qquad c_{jk} = \bar{c}_{-jk}$$

and let L be the operator

$$u \mapsto \sum_{j=-\infty}^{\infty} \sum_{k=1}^{\infty} (k^2-j^2) c_{jk} \sin kx e^{ijt}$$

with domain

$$\mathcal{D}(L) = \{u \in L^2((0,\pi) \times [0,2\pi]) \mid \sum |k^2-j^2|^2 |c_{jk}|^2 < +\infty\}.$$

If $k^2 - j^2 \neq \gamma$, $k = 0,1,\ldots,$ $j = 0,\pm1,\pm2,\ldots,$ then

$$\frac{k+|j|}{|k^2-j^2-\gamma|} \leq \text{const. } C.$$

It follows that

$$\|u\|^2_{H^{\frac{1}{2}}} = \sum (k+|j|)|c_{jk}|^2 \leq C \sum |k^2-j^2-\gamma|\,|c_{jk}|^2$$
$$= C\| |L-\gamma I|^{\frac{1}{2}} u\|^2.$$

Now, the Hölder inequality (9.8) holds for

$$\frac{1}{\gamma} + \frac{2}{p} = 1, \quad p = \frac{4n}{n-4}.$$

Again, $\Phi \in C^2(D(|L - \gamma I|^{\frac{1}{2}}), \mathbb{R}^1)$ is verified.

COROLLARY 9.2. *Suppose that* $g(x,t,\xi)$ *is a* C^1 *function defined on* $[0,\pi] \times [0,2\pi] \times \mathbb{R}^1$, *satisfying the following conditions:*

(1) $\lim_{|\xi|\to\infty} \frac{g(x,t,\xi)}{\xi} = \lambda_\infty \notin \sigma(A)$ *uniformly in* $(x,t) \in [0,\pi] \times [0,2\pi]$;

(2) $g(x,t,0) \equiv 0$;

(3) *there exist* $\alpha < \beta$ *with either* $0 < \alpha$ *or* $\beta < 0$, *such that*

$$\alpha \leq g'_\xi(x,t,\xi) \leq \beta;$$

(4) *either* $g'_\xi(x,0) \le \bar\lambda - \varepsilon < \bar\lambda < \lambda_\infty$, *or* $g'_\xi(x,0) \ge \bar\lambda + \varepsilon > \bar\lambda > \lambda_\infty$, *where* $\bar\lambda \in \sigma(A)$, *and* A *is the self-adjoint wave operator with Dirichlet-*2π*-periodic condition.*

Then the equation (9.1') has at least one nontrivial solution.

9.3. Super- and sub-solutions and local minimum

In the following, we shall say that $\underline{u} \in C^2(\Omega)$ (or $\bar u$) is a *sub-* (or *super-* resp.) *solution* of the following equation:

$$(9.1) \qquad \begin{cases} -\Delta u = g(x,u), & x \in \Omega, \\ u|_{\partial\Omega} = 0, \end{cases}$$

if

$$\begin{cases} -\Delta \underline{u} < g(x,\underline{u}(x)) & \text{in } \Omega, \\ \underline{u}|_{\partial\Omega} = 0 \end{cases} \qquad \left(\text{or } \begin{cases} -\Delta \bar u \ge g(x,\bar u(x)) & \text{in } \Omega \\ \bar u|_{\partial\Omega} = 0 \end{cases} \right).$$

If $\underline{u} < \bar u$, we write $C = \{u \in H^1_0(\Omega) \mid \underline{u}(x) \le u(x) \le \bar u(x) \text{ a.e.}\}$, and $C_X = C \cap C^1_0(\bar\Omega)$.

It is well known cf. Amann [1] that if there is a pair of sub- and super-solutions $\underline{u} < \bar u$ of (9.1), then there is a solution $u_0 \in C$ of (9.1). One asks can we characterize the solution by the corresponding functional

$$f(u) = \int_\Omega \left[\frac{(\nabla u)^2}{2} - G(x,u)\right] dx \; ?$$

Now we shall prove that f is bounded from below on C_X, and then attains its minimum, which is the variational characterization of u_0. Applying Example 1 in Section 1, we obtain the critical groups of u_0:

$$(9.10) \qquad C_k(f,u_0) = \begin{cases} G & k = 0 \\ 0 & k \ne 0. \end{cases}$$

THEOREM 9.3. *Suppose that* $\underline{u} < \bar u$ *is a pair of sub- and super-solutions of* (9.1), *then there is a point* $u_0 \in C_X$ *which is a local minimum of the functional* $\tilde{f} = f|_{C_0^1(\bar{\Omega})}$, *so that*

$$C_k(f,u_0) = \begin{cases} G & k = 0 \\ 0 & k \neq 0. \end{cases} \tag{9.11}$$

PROOF. One may assume that $\underline{u}(x) < 0 < \bar{u}(x)$, without loss of generality. Define a new function

$$\hat{g}(x,\xi) = \begin{cases} g(x,\bar{u}(x) \wedge (-\Delta\bar{u}(x)) , & \xi > \bar{u}(x) \\ g(x,\xi) , & \underline{u}(x) \le \xi \le \bar{u}(x) \\ g(x,\underline{u}(x)) \vee (-\Delta\underline{u}(x)), & \xi < \underline{u}(x) \end{cases} \tag{9.12}$$

where $a \vee b = \max\{a,b\}$, and $a \wedge b = \min\{a,b\}$. By definition, $\hat{g}(x,\xi) \in C(\bar{\Omega}\times\mathbf{R}^1)$, is bounded, and satisfies:

$$g(x,\xi) = \hat{g}(x,\xi) \quad \text{for} \quad \underline{u}(x) \le \xi \le \bar{u}(x).$$

Let

$$\hat{G}(x,\xi) = \int_0^\xi \hat{g}(x,t)\, dt.$$

Then $\hat{G} \in C^1(\bar{\Omega}\times\mathbf{R}^1)$, and the functional

$$J(u) = \int_\Omega \left[\frac{|\nabla u|^2}{2} - \hat{G}(x,u)\right] dx$$

defined on $H_0^1(\Omega)$ is bounded from below, and satisfies the P.S. condition. Hence there is a minimum u_0 which satisfies

$$dJ(u_0) = \theta$$

i.e., u_0 satisfies the equation

$$\begin{cases} -\Delta u_0 = \hat{g}(x,u_0) \\ u_0|_{\partial\Omega} = 0. \end{cases}$$

According to the L^p regularity of solutions of elliptic BVP, and the maximum principle, we see that $u_0 \in C_X$. But

$$\tilde{f} = f|_{C_X} = J|_{C_X},$$

therefore u_o is a local minimum of $\tilde{f}$, which implies (9.11).

Whenever f is well-defined on $H_o^1(\Omega)$, we obtain (9.10).

REMARK 9.3. Theorem 9.3 was first proved in Chang [23] by an abstract method.

9.4. A bifurcation problem

For simplicity, the function $g(x,\xi)$ in (9.1) is replaced by $g(\xi)$. We assume that

(1) $\lim_{|\xi|\to\infty} \frac{g(\xi)}{\xi} < \lambda_1$, the first eigenvalue of $-\Delta$ with 0-Dirichlet boundary value.

(2) $g(0) = 0$, and $g \in C^1(\mathbb{R}^1)$.

THEOREM 9.4. *Let* $\lambda = g'(0)$, *then*

(i) *For* $\lambda > \lambda_1$, *the* BVP (9.1) *has at least two nontrivial solutions.*

(ii) *For* $\lambda > \lambda_2$, *with* $\lambda \notin \sigma(-\Delta)$, (9.1) *has at least four nontrivial solutions.*

(iii) *For* $\lambda > \lambda_2$, *with* $\lambda \in \sigma(-\Delta)$, *if either*

$$\frac{g(\xi)}{\xi} > \lambda \quad \text{or} \quad \frac{g(\xi)}{\xi} < \lambda \tag{9.13}$$

for $\xi \neq 0$ *in a neighbourhood* U *of* 0, *then the same conclusion as in* (ii) *holds.*

PROOF. By condition (1), there exists an $\alpha \in (0,\lambda_1)$ and a constant $C_\alpha > 0$ such that $g(\xi) \leq \alpha\xi + C_\alpha$ if $\xi > 0$, $g(\xi) \geq -\alpha\xi - C_\alpha$. Let φ_o be the solution of the following equation

$$\begin{cases} -\Delta\varphi_o = \alpha\varphi_o + C_\alpha & \text{in } \Omega \\ \varphi_o|_{\partial\Omega} = 0 & . \end{cases}$$

Then, by the maximum principle, $\varphi_o > 0$, and hence, $-\varphi_o < \varphi_o$ is a pair of sub-

and super-solutions of (9.1).

According to the cut-off technique in (9.12), we may assume that $g(\xi)$ is bounded, and define the following functional

$$f(u) = \int_\Omega \left[\frac{(\nabla u)^2}{2} - G(u)\right] dx \quad \text{on} \quad H_0^1(\Omega),$$

which is bounded from below, so that the P.S. condition is satisfied.

(i) Let φ_1 be the first eigenfunction, with $\max_{x\in\bar\Omega} \varphi_1 = 1$, and $\varphi_1 > 0$. We may choose $\varepsilon > 0$ so small that

$$-\varphi_o < -\varepsilon\varphi_1 \quad \text{and} \quad \varepsilon\varphi_1 < \varphi_o$$

are two pairs of sub- and super-solutions of (9.1). According to Lemma 9.1, we have two distinct solutions $z_1, z_2 \in H_0^1(\Omega)$, satisfying

$$C_k(f,z_i) = \begin{cases} G & k = 0 \\ 0 & k \neq 0 \end{cases}, \quad i = 1,2,$$

if they are isolated.

(ii) We may assume that there are at most finitely many solutions. Lemma 7.1, or its variant, Remark 7.3, is employed. We obtain a third solution z_3. According to a result due to Hess and Kato [41] the condition (7.3) is fulfilled. On account of Theorem 7.1, we have

$$C_k(f,z_3) = \begin{cases} G & k = 1 \\ 0 & k \neq 1. \end{cases}$$

We shall prove that $z_3 \neq \theta$. In fact,

$$d^2f(\theta) = id - \lambda(-\Delta)^{-1}$$

but the condition $\lambda > \lambda_2$, with $\lambda \notin \sigma(-\Delta)$, means that θ is a nondegenerate critical point with index $j \geq 1 + k_2$, where $k_2 = \dim \ker(id - \lambda_2(-\Delta)^{-1})$. From Theorem 1.1 we have

$$C_k(f,\theta) = \begin{cases} G & k = j \\ 0 & k \neq j. \end{cases}$$

The critical groups isolate z_3 from θ.

If there were no other critical points, then a contradiction would happen provided by the Morse inequalities: $\beta_o = 1$, $\beta_k = 0$, $k \neq 0$. In fact, for $k > j$, one would have

$$M_k - M_{k-1} + \dots + (-1)^k M_o = \beta_k - \beta_{k-1} + \dots + (-1)^k \beta_o.$$

The LHS is even, but the RHS is odd. Therefore there are at least four nontrivial solutions.

(iii) In case $\lambda > \lambda_2$, but $\lambda \in \sigma(-\Delta)$, θ is a degenerate critical point. The assumption $g(\xi)/\xi > \lambda$ for $\lambda \in U\backslash\{0\}$ implies that $\tilde{f} = f|_N \leq 0$, where N is a neighbourhood in the characteristic submanifold at θ, $\dim N = \dim \ker(-\Delta-\lambda I)$, say, equals to m_o. Let m_- be the Morse index of f at θ. We have $m_- \geq 2$, and

$$C_k(f,\theta) = \begin{cases} G & k = m_- + m_o \\ 0 & k \neq m_- + m_o \end{cases}$$

because θ is a local maximum of $\tilde{f}$, Theorem 2.4 is employed. The same argument used in (ii) gives the fourth nontrivial solution.

Similarly, in case $g(\xi)/\xi < \lambda$ for $\xi \in U\backslash\{0\}$, θ is a local minimum of $\tilde{f}$; thus

$$C_k(f,\theta) = \begin{cases} G & k = m_- \\ 0 & k \neq m_-. \end{cases}$$

The theorem is proved.

A special form of this problem is that the function is of the following form:

$$g(u) = \lambda u - h(u) \tag{9.14}$$

where λ is a real parameter and $h(u)$ satisfies the following conditions: $h \in C^1(\mathbb{R}^1)$, $h(0) = h'(0) = 0$ and $\lim\limits_{|u|\to+\infty} \frac{h(u)}{u} = +\infty$.

In this sense, we call it a bifurcation problem.

REMARK 9.4. The special form (9.14) of Theorem 9.4 has been studied by many authors. Cf. Ambrosetti [7], Struwe [77], for at least three solutions, Hofer [42], Tian [79] for at least four solutions in cases (i) and (ii). The result in case (iii) was independently obtained by Dancer [32].

9.5. Jumping nonlinearities

Elliptic equations with jumping nonlinearities were first studied by Ambrosetti and Prodi [9], and followed by many authors: cf. Amann and Hess [4], Berger and Podolak [16], Fučik [36], Kazdan and Warner [44], Hess [40], Dancer [32], H. Berestycki and P.L. Lions [14] and the author [24]. After an observation due to Lazer and McKenna [49], more solutions were obtained. In this respect, the reader is referred to Solimini [76], Ambrosetti [8], Hofer [42] and Dancer [32].

We consider the following BVP with a real parameter $t \in \mathbb{R}^1$.

$$(P_t)\begin{cases} -\Delta u = f(x,u) + t\varphi_1 & \text{in } \Omega \\ u|_{\partial\Omega} = 0 \end{cases} \tag{9.15}$$

where φ_1 is the first eigenfunction, with $\varphi_1(x) > 0$ $\forall x \in \Omega$. Assume that $f \in C^1(\overline{\Omega}\times\mathbb{R}^1)$, satisfying the following conditions:

(1) $\lim\limits_{\xi\to+\infty} f'_\xi(x,\xi) = \gamma$ uniformly in $x \in \overline{\Omega}$, and $\gamma \in (\lambda_j,\lambda_{j+1})$ for some $j \geq 1$, where $\{\lambda_j \mid j = 1,2,\ldots\} = \sigma(-\Delta)$.

(2) $\overline{\lim\limits_{\xi\to-\infty}} \frac{f(x,\xi)}{\xi} \leq \lambda_1 - \delta$, uniformly in $x \in \overline{\Omega}$, for some $\delta > 0$.

(3) There exists a constant M such that

$$|f'_{\xi}(x,\xi)| \le M(1 + |\xi|^{\frac{4}{n-2}}).$$

We note that condition (1) implies that

$$\lim_{\xi\to+\infty} \frac{f(x,\xi)}{\xi} = \gamma.$$

THEOREM 9.5. *Suppose that the conditions* (1)-(3) *are fulfilled. Then there exists* $t^* \in \mathbb{R}^1$ *such that* (P_t) *has*

(1) *no solution, if* $t > t^*$;

(2) *at least one solution, if* $t = t^*$;

(3) *at least two solutions, if* $t < t^*$.

If further, we assume $j \ge 2$, *i.e.,* $\gamma > \lambda_2$, *the second eigenvalue of* $-\Delta$, *then there exists* $t^{**} < t^*$ *such that* (P_t) *has at least four solutions if* $t < t^{**}$.

The proof depends on the following lemmas.

LEMMA 9.1. *Assume conditions* (1), (2), *and* $f \in C(\bar{\Omega}\times\mathbb{R}^1)$. *Let*

$$J_t(u) = \int_\Omega \left[\frac{|\nabla u|^2}{2} - F(x,u) - t\varphi_1 u\right] dx \qquad u \in H^1_0(\Omega)$$

where $F(x,\xi) = \int_0^\xi f(x,s)\, ds$. *Then for all* $t \in \mathbb{R}^1$, J_t *satisfies the P.S. condition.*

PROOF. For each function $u \in L^1_{loc}(\Omega)$ we denote $u^+ = \max\{u,0\}$, and $u^- = u - u^+$. Assume that $\{u_n\} \subset H^1_0(\Omega)$ is a sequence satisfying

$$\int_\Omega (\nabla u_n \cdot \nabla v - f(x,u_n)v - t\varphi_1 v)\, dx = o(\|v\|) \qquad v \in H^1_0(\Omega) \tag{9.16}$$

where $\|.\|$ is the $H^1_0(\Omega)$ norm. Then we obtain

$$\int_\Omega [\nabla u_n^+ \nabla v - f(x,u_n)^+ v - (t\varphi_1)^+ v]\, dx = o(\|v\|) \qquad \forall v \in H^1_0(\Omega).$$

Let $\rho_n = \rho_n^+ + \rho_n^-$, where

$$\rho_n^\pm = u_n^\pm - (-\Delta)^{-1}[f(x,u_n)^\pm + (t\varphi_1)^\pm] \to \theta \ \text{ in } H^1_0(\Omega).$$

By the condition (1), $\|f(x,u_n)^+ - \gamma u_n^+\|_{L^2} = o(\|u_n^+\|_{L^2})$; but

$$u_n^+ = (id - \gamma(-\Delta)^{-1})^{-1}\{(-\Delta)^{-1}[(f(x,u_n)^+ - \gamma u_n^+) + t^+\varphi_1] + \rho_n^+\};$$

it follows that $\{\|u_n^+\|\}$ is bounded. From conditions (1) and (2), we have $\delta > 0$ and $C > 0$ such that

$$f(x,\xi) - \lambda_1\xi > \delta|\xi| - C. \tag{9.17}$$

Let us choose $\mu < \lambda_1$ such that $\lambda_1 - \mu < \delta$, then we have

$$\begin{aligned}(-\Delta_1-\mu)(u_n-\rho_n) &= f(x,u_n) + t\varphi_1 - \mu(u_n-\rho_n)\\ &\geq -C + t\varphi_1 + \mu\rho_n.\end{aligned}$$

By the weak Maximum Principle, one deduces

$$u_n - p_n \geq (-\Delta-\mu)^{-1}[-C + t\varphi_1 + \mu\rho_n] \tag{9.18}$$

noticing $\rho_n \to \theta$ $(H_o^1(\Omega))$. Combining (9.18) with the boundedness of u_n^+, we obtain that

$$\|u_n\|_{L^2} \text{ is bounded.}$$

Substituting it into (9.16), $\{\|u_n\|\}$ is bounded. After a standard procedure, the P.S. condition is verified.

LEMMA 9.2. *Under the conditions* (1) *and* (2) *there exists a subsolution* $\underline{u}_t$ *for the* BVP (P_t) *such that for each solution* u_t *of* (P_t) *we have* $u_t > \underline{u}_t$.

PROOF. According to (9.18), if we define $\underline{u}_t$ to be the solution of the following BVP:

$$\begin{cases} -\Delta u - \mu u = -C + t\varphi_1 & \text{in } \Omega \\ u|_{\partial\Omega} = 0 \end{cases}$$

the conclusion follows from the weak Maximum Principle, and the inequality (9.17).

LEMMA 9.3. *Under the conditions* (1), (2) *there exists* $t_o \in \mathbb{R}^1$ *such that if* (P_t) *is solvable, then* $t < t_o$.

PROOF. By (9.17), we have $\delta > 0$, $C > 0$ such that

$$f(x,\xi) - \lambda_1\xi \geq \delta|\xi| - C.$$

Thus, if u_t is a solution of (P_t), then multiplying by φ_1 on both sides of the equation, and by integration, we obtain

$$\lambda_1 \int_\Omega u_t\varphi_1 \, dx = \int_\Omega f(x,u_t)\varphi_1 \, dx + t\int_\Omega \varphi_1^2 \, dx.$$

From this one deduces

$$t\int_\Omega \varphi_1^2 \, dx + \delta\int_\Omega |u_t|\cdot\varphi_1 \, dx - C\int_\Omega \varphi_1 \, dx \leq 0,$$

that is

$$t < \left(\int_\Omega \varphi_1^2 \, dx\right)^{-1} C\int_\Omega \varphi_1 \, dx.$$

LEMMA 9.4. *Under the conditions* (1), (2) *and* (3), *there exists* $t_1 \in \mathbb{R}^1$ *such that* (P_t) *possesses a positive solution* $\bar{u}_t$ *which is a nondegenerate critical point of* J_t *with index* $d^2J_t(\bar{u}_t) = h_j$ *for* $t < t_1$, *where*
$h_j = \sum_{k\leq j} \dim\ker(-\Delta - \lambda_k I)$.

PROOF. Let

$$g(x,\xi) = -\gamma\xi + f(x,\xi) \qquad \xi \geq 0.$$

We have $g(x,\xi) = o(|\xi|)$ uniformly in $x \in \bar{\Omega}$, as $\xi \to +\infty$. We extend the function g to be a function $\hat{g}$ such that $\hat{g} \in C^1(\bar{\Omega}\times\mathbb{R}^1)$, with $|\hat{g}(x,\xi)| = o(|\xi|)$ uniformly in $x \in \bar{\Omega}$. According to Theorem 9.1, the equation

$$\begin{cases} -\Delta u = \gamma u + \hat{g}(x,u) + t\varphi_1 & \text{in } \Omega \\ u|_{\partial\Omega} = 0 \end{cases}$$

possesses a solution $\bar{u}_t$. Define

$$v_t = \bar{u}_t - \frac{t\varphi_1}{\lambda_1 - \gamma}. \tag{9.19}$$

We obtain

$$\begin{cases} -\Delta v_t = \gamma v_t + \hat{g}(x,\bar{u}_t) & \text{in } \Omega, \\ v_t|_{\partial\Omega} = 0. \end{cases}$$

Thus, the L^p a priori bounds for v_t are employed to deduce

$$\|v_t\|_{C^1} = o(|t|) \quad \text{as} \quad |t| \to \infty.$$

Substituting the estimate into (9.19), we obtain

$$\bar{u}_t > 0 \quad \text{for} \quad t < t_1$$

where $-t_1$ is a real number large enough. This proves that $\bar{u}_t$ is a solution of the problem (P_t).

Again, we have

$$\bar{u}_t(x) = v_t(x) + \frac{t}{\lambda_1 - \gamma}\varphi_1(x) \to +\infty, \text{ a.e.} \quad \text{as} \quad t \to -\infty.$$

This implies that

$$f'_\xi(x,\bar{u}_t(x)) \to \gamma \quad \text{a.e. as} \quad t \to -\infty.$$

Since

$$d^2J(\bar{u}_t) = id - (-\Delta)^{-1} f'_\xi(x,\bar{u}_t(x)),$$

by the Hölder inequality as well as the Sobolev embedding theorem, we see that

$$\|d^2J(\bar{u}_t) - (id - \gamma(-\Delta)^{-1})\|_{L(H^1_0(\Omega))} \le \left(\int_\Omega |f'_\xi(x,\bar{u}_t(x)) - \gamma|^p \, dx\right)^{\frac{1}{p}},$$

where $p = \frac{n}{2}$. Applying the Lebesgue dominance theorem, we arrive at

$$\|d^2J(\bar{u}_t) - (id - \gamma(-\Delta)^{-1})\|_{L(H^1_0(\Omega))} = o(1).$$

Because $\gamma \notin \sigma(-\Delta)$ is assumed, $id - \gamma(-\Delta)^{-1}$ is invertible. Thus $\bar{u}_t$ is non-degenerate with Morse index h_j.

PROOF of Theorem 9.5. Define $t^* = \sup\{t \in \mathbb{R}^1 \mid (P_t) \text{ is solvable}\}$. Combining Lemma 9.3 with Lemma 9.4, $t^* \in \mathbb{R}^1$, hence (1) is proved.

Now we come to look for a strict supersolution for $t < t^*$. In fact, choose $t' \in (t,t^*)$, let $u_{t'}$ be a solution of $(P_{t'})$, then $u_{t'}$ is a strict supersolution of (P_t) which satisfies $u_{t'} > \underline{u}_{t'} > \underline{u}_t$. Hence $[\underline{u}_t, u_{t'}]$ is a pair of strict sub- and supersolutions of (P_t). Theorem 9.3 is employed to deduce a solution u_t of (P_t) which is a local minimum of the functional $\tilde{J}_t = J_t|_{C_0^1(\bar{\Omega})}$, so that

$$C_k(J_t,u_t) = \begin{cases} G & k = 0, \\ 0 & k \neq 0. \end{cases}$$

Noticing that the functional $\tilde{J}_t$ is unbounded from below along the ray $u_s = s\varphi_1$, $s > 0$, the generalized Mountain Pass lemma can be applied (Theorem 7.1, and Lemma 7.1). In combining this with a result due to Hess and Kato mentioned before, we find a second solution $\hat{u}_t$ with critical groups

$$C_k(J_t,\hat{u}_t) = \begin{cases} G & k = 1, \\ 0 & k \neq 1. \end{cases}$$

The conclusion (3) is proved.

As to the conclusion (2), we prove by the same method as in Lemma 9.1, that the set $\{u_t \mid t \in [t^*-1,t^*]\}$, where u_t is the solution of (P_t) obtained by the previous sub- and supersolutions is bounded in $H_0^1(\Omega)$. We obtain a sequence $t_i \to t^*$ such that u_{t_i} weakly converges in $H_0^1(\Omega)$, say to u^*. Then u^* is a solution of (P_{t^*}).

Finally, we assume $\gamma \in (\lambda_j,\lambda_{j+1})$, with $j \geq 2$. According to Lemma 9.4, there is a $t^{**} < t^*$ such that there exists a third solution $\bar{u}_t$ of (P_t) such that $\bar{u}_t$ is nondegenerate, with

$$C_k(J_t,\bar{u}_t) = \begin{cases} G & k = h_j, \\ 0 & k \neq h_j. \end{cases}$$

One more solution will then be obtained by a computation of the Leray-

Schauder degree. In fact, by Lemma 9.1, we conclude that all solutions of the equation

$$u = (-\Delta)^{-1}(f(x,u) + t\varphi_1) \tag{9.20}$$

are bounded in an open ball $\overset{\circ}{B}_{R_t}$, where R_t, the radius, depends on t continuously. By the homotopy invariance of the Leray-Schauder degree, one has

$$\deg(\mathrm{id}-(-\Delta)^{-1}F_t, \overset{\circ}{B}_{R_t}, \theta) = \text{const.} \qquad \forall t \in \mathbb{R}^1,$$

where

$$F_t u = f(x,u(x)) + t\varphi_1(x).$$

But, from conclusion (1), if $t > t^*$, (9.20) has no solution. It follows that

$$\deg(\mathrm{id}-(-\Delta)^{-1}F_t, \overset{\circ}{B}_t, \theta) = 0, \qquad \forall t \in \mathbb{R}^1. \tag{9.21}$$

If $t < t^{**}$, suppose that there are only three solutions $\bar{u}_t$, $\hat{u}_t$ and u_t, then by Theorem 4.1 the Leray-Schauder degree would be

$$\deg(\mathrm{id}-(\Delta)^{-1}F_t, \overset{\circ}{B}_t, \theta) = (-1)^{h_j}. \tag{9.22}$$

This will be a contradiction.

REMARK 9.5. Lemmas 9.2 and 9.3 are due to Kazdan and Warner [44], and Lemma 9.4 is due to Ambrosetti [8], Lazer and McKenna [49]. The idea of the proof is taken from Hofer [42], Dancer [32] and Chang [24].

9.6. Other examples

Suppose that $g \in C^1(\mathbb{R}^1)$, satisfies the following conditions

(1) $g(0) = 0$, $0 \leqq g'(0) < \lambda_1$;

(2) $g'(t) > 0$ and strictly increasing in t for $t > 0$;

(3) $g'(\infty) = \lim\limits_{|t|\to\infty} g'(t)$ exists and lies in (λ_1, λ_2).

THEOREM 9.6. *Under the conditions* (1), (2), (3), *the equation*

(9.23) $$\begin{cases} -\Delta u = g(u) & \textit{in } \Omega \\ u|_{\partial\Omega} = 0 \end{cases}$$

has at least three distinct solutions.

PROOF. 1° It is obvious that θ is a solution, which is also a strict local minimum of the functional:

$$J(u) = \int_\Omega [\tfrac{1}{2}(\nabla u)^2 - G(u)]\, dx \quad \text{on} \quad H_0^1(\Omega)$$

where G is the primitive of g, with $G(0) = 0$.

2° Modify g to be a new function

$$\hat{g}(\xi) = \begin{cases} g(\xi) & \xi \geq 0 \\ 0 & \xi = 0 \end{cases}$$

and consider a new functional

$$\hat{J}(u) = \int_\Omega [\tfrac{1}{2}(\nabla u)^2 - \hat{G}(u)]\, dx,$$

where $\hat{G}(\xi) = \int_0^\xi \hat{g}(t)\, dt$. It is easily seen that θ is also a strict local minimum of $\hat{J}$, which is a C^1 functional with P.S. condition.

Since $\hat{J}$ is unbounded from below, along the ray $u_s = s\varphi_1(x)$, $s > 0$, the generalized mountain pass lemma (Lemma 7.1) yields a critical point u_0 of $\hat{J}$ which solves the equation

$$\begin{cases} -\Delta u = \hat{g}(u) & x \in \Omega, \\ u|_{\partial\Omega} = 0. \end{cases}$$

Since $\hat{g}(u) \geq 0$, by the Maximum Principle, $u_0 \geq 0$, hence u_0 is a solution of (9.23).

3° Now we shall prove that $-\Delta - g'(u_0(x))$ is an invertible operator defined on $L^2(\Omega)$.

Since u_0 satisfies (9.23), it is also a solution of the equation

$$-\Delta u_o - q(x)u_o(x) = 0, \quad u_o|_{\partial\Omega} = 0,$$

where

$$q(x) = \int_0^1 g'(tu_o(x))\, dt.$$

Let $\mu_1 < \mu_2 < \ldots$ be the eigenvalues of the problem

$$\begin{cases} -\Delta w - \mu g'(u_o(x))w = 0, \ g \\ w|_{\partial\Omega} = 0. \end{cases}$$

We shall prove that $\mu_1 < 1 < \mu_2$. This implies the invertibility of the operator $-\Delta - g'(u_o(x))$.

In fact, according to assumption (2), we have

$$q(x) < g'(u_o(x)) \qquad \forall x \in \Omega$$

so that

$$\mu_1 = \min \frac{\int (\nabla w)^2}{\int g'(u_o)w^2} < \min \frac{\int (\nabla w)^2}{\int q(x)w^2} \le 1.$$

Again, by assumptions (2) and (3), we have

$$g'(u_o(x)) < \lambda_2 \qquad \forall x \in \Omega.$$

According to the Rayleigh quotient characterization of the eigenvalues

$$\mu_2 = \sup_{E_1} \inf_{w \in E_1^\perp} \frac{\int (\nabla w)^2}{\int g'(u_o(x))w^2} > \frac{1}{\lambda_2} \sup_{E_1} \inf_{w \in E_1^\perp} \frac{\int (\nabla w)^2}{\int w^2} = 1$$

where E_1 is any one dimensional subspace in $H_o^1(\Omega)$.

4° Now we use the finite dimensional reduction for the equation (9.23), which is regarded to be an operator equation defined on $L^2(\Omega)$. By Theorem 8.4, it is reduced to finding the critical points of a function $a \in C^2(\mathbb{R}^N, \mathbb{R}^1)$, where N is an integer depending on the range of g'. Noticing that

$$d^2a(z) = (-\Delta - g'(u(z)))u'(z), \qquad z \in \mathbb{R}^N$$

where $z \mapsto u(z)$ is the correspondence of $z \in \mathbb{R}^N$ to a function $u \in L^2(\Omega)$.

If z_o corresponds to $u_o(x)$, then we shall prove that z_o is a nondegenerate critical point of $a(z)$. In fact, suppose that for some $w \in \mathbb{R}^N$ we have $d^2a(z_o)w = \theta$. Then

$$(-\Delta - g'(u_o(x)))u'(z_o)w = \theta$$

which implies that

$$u'(z_o)w = \theta,$$

because $-\Delta - g'(u_o(x))$ is invertible. This yields $w = \theta$.

The Morse identity (1.7) yields an odd number of critical points. Therefore there are at least three solutions of (9.23) (cf. Corollary 8.2).

Finally, we turn to the following example.

THEOREM 9.7. *Suppose that* $g \in C^1(\mathbb{R}^1)$ *satisfies the following conditions:*

(1) $g(0) = 0$, *and* $\lambda_2 < g'(0) < \lambda_3$;

(2) $g'(\infty) = \lim_{t\to\pm\infty} g'(t)$ *exists, and* $g'(\infty) \notin \sigma(-\Delta)$, *with* $g'(\infty) > \lambda_3$;

(3) $|g(t)| < 1$ *and* $0 \leq g'(t) < \lambda_3$ *in the interval* $[-c,c]$, *where* $c = \max_{x\in\bar\Omega} e(x)$, *and* $e(x)$ *is the solution of the* BVP:

$$\begin{cases} -\Delta e = 1 & \text{in } \Omega \\ e|_{\partial\Omega} = 0. \end{cases}$$

Then the equation (9.23) *possesses at least five nontrivial solutions.*

PROOF. Define

$$(9.24) \qquad \tilde g(t) = \begin{cases} g(c) & \text{if } t > c \\ g(t) & \text{if } |t| \leq c \\ g(-c) & \text{if } t < -c \end{cases}$$

and let

$$\tilde{J}(u) = \int_\Omega [\tfrac{1}{2}(\nabla u)^2 - \tilde{G}(u)]\, dx,$$

where $\tilde{G}(t) = \int_0^t \tilde{g}(s)\, ds$. The truncated equation

$$\begin{cases} -\Delta u = \tilde{g}(u) & \text{in } \Omega \\ u|_{\partial\Omega} = 0 \end{cases} \tag{9.25}$$

possesses at least three solutions θ, u_1, u_2, because there are two pairs of sub- and supersolutions $[\varepsilon\varphi_1, e]$ and $[-e, -\varepsilon\varphi_1]$, where φ_1 is the first eigenfunction of $-\Delta$, with $\varphi_1(x) > 0$, and $\varepsilon > 0$ a small enough constant.

By the generalized Mountain Pass lemma, there is a mountain pass point u_3. That $u_3 \neq \theta$ follows from the fact that

$$C_k(\tilde{J}, u_3) = \begin{cases} G & k = 1 \\ 0 & k \neq 1. \end{cases}$$

But from condition (1)

$$C_k(\tilde{J}, \theta) = \begin{cases} G & k = m_1 + m_2 \\ 0 & k \neq m_1 + m_2, \end{cases}$$

where $m_i = \dim \ker(-\Delta - \lambda_i I)$, $i = 1,2,\ldots$ By Theorem 9.3, one has

$$C_k(\tilde{J}, u_i) = \begin{cases} G & k = 0 \\ 0 & k \neq 0 \end{cases}, \quad i = 1,2.$$

Noticing that $\tilde{J}$ is bounded from below, we conclude that there is at least another critical point u_4.

Obviously, all these critical points u_i, $i = 1,2,3,4$, are solutions of the equation (9.25).

On account of the first condition in (3), in combination with the Maximum Principle, all solutions of (9.25) are bounded in the interval $[-c, c]$. Therefore they are solutions of (9.24).

Next, a finite dimensional reduction as in Section 8 is applied. Let us denote $z \mapsto u$ the correspondence, and let a be the function defined on $\mathbb{R}^N$ whose critical points are solutions of the equation (9.24). If u is a solution, by the second condition in (3),

$$\mathrm{ind}(d^2a(z)) + \dim \ker(d^2a(z)) \leq m_1 + m_2,$$

because the negative and zero invariant subspaces of the operator $-\Delta - g'(u(x))$ are included in $\bigoplus_{j=1}^{2} (-\Delta-\lambda_j I)$. But

$$d^2a(\infty) = -\Delta - g'(\infty)I\Big|_{\mathbb{R}^N}$$

so we have

$$\mathrm{ind}(d^2a(\infty)) > m_1 + m_2.$$

By Theorem 8.2, there exists another critical point z_5 of a, which yields the fifth nontrivial solution of the equation (9.24).

Section 10

MULTIPLE PERIODIC SOLUTIONS OF HAMILTONIAN SYSTEMS

In the last section of these lectures, we shall apply the Morse theory to estimate numbers of periodic solutions of Hamiltonian systems.

Let $H(t,x)$ be a C^2 function defined on $\mathbb{R}^1 \times \mathbb{R}^{2n}$ which is 1-periodic with respect to the first variable t. We are interested in the existence and multiplicity of the 1-periodic solutions of the following Hamiltonian system:

$$\text{(10.1)} \qquad \begin{cases} \dot{q} = -H_p(t;q,p) \\ \dot{p} = H_q(t;q,p) \end{cases}$$

where $q,p \in \mathbb{R}^n$, $x = (q,p)$. The function H then is called the Hamiltonian function. Letting J be the standard symplectic structure on $\mathbb{R}^{2n}$, i.e.,

$$J = \begin{pmatrix} 0 & -I_n \\ I_n & 0 \end{pmatrix}$$

where I_n is the $n \times n$ identity matrix, the equation (10.1) can be written in a compact version

$$\text{(10.2)} \qquad -J\dot{x} = H_x(t,x).$$

The equation (10.2) is a special case of the operator equation considered in Section 8. In fact, let $X = L^2((0,1),\mathbb{R}^{2n})$ and let

$$A: x(t) \mapsto -J\dot{x}(t), \tag{10.3}$$

with domain

$$D(A) = \hat{H}^1((0,1),\mathbb{R}^{2n}) = \{x(t) \in H^1((0,1),\mathbb{R}^{2n}),\ x(0) = x(1)\}$$

and define a functional

$$\Phi(x) = \int_0^1 H(t,x(t))\,dt. \tag{10.4}$$

Suppose that there is a constant $C > 0$ such that

$$\|\partial_x^2 H(t,x)\|_{L(\mathbb{R}^{2n})} \le C \tag{10.5}$$

then $\Phi \in C^1(X,\mathbb{R}^1)$, with

$$F(x) \triangleq \Phi'(x) = H_x(t,x(t)) \tag{10.6}$$

and then F is Gateaux differentiable, with

$$dF(x)y = \partial_x^2 H(t,x)y \qquad \forall y \in X. \tag{10.7}$$

The equation (10.2) is of the form

$$Ax = F(x). \tag{10.8}$$

It is easily verified that

(1) A is a self-adjoint operator with spectrum $\sigma(A) = 2\pi\mathbb{Z}$. Let $\{e_1,\dots,e_{2n}\}$ be an orthonormal basis in $\mathbb{R}^{2n}$, then $\forall j \in \mathbb{Z}$

$$M(j) = \text{span}\{\cos 2\pi jt e_k + \sin 2\pi jt Je_k,\ k = 1,2,\dots,2n\}$$

is the eigenspace corresponding to $2\pi j \in \sigma(A)$. We may assume that $C \notin 2\pi\mathbb{Z}$, in (10.5), and let $-\alpha = \beta = C$, then the assumption (A) in Section 8 is satisfied.

(2) The condition (10.5) implies (10.7) and that

$$\|dF\|_{L(X)} \le C.$$

This is assumption (F) in Section 8.

(3) Observing the following continuous imbeddings:

$$\mathcal{D}(|A|^{\frac{1}{2}}) \to H^{\frac{1}{2}}((0,1),\mathbb{R}^{2n}) \to L^p((0,1),\mathbb{R}^{2n}) \qquad \forall p < \infty,$$

condition (10.5) implies that $\Phi \in C^2(\mathcal{D}(|A|^{\frac{1}{2}}),\mathbb{R}^1)$, which is assumption (D) in Section 8.

10.1. A nonresonance problem

Firstly, we assume that there exists a symmetric $2n \times 2n$ matrix B_∞ such that

(10.9) $$\|H_x(t,x) - B_\infty x\|_{\mathbb{R}^{2n}} = o(\|x\|_{\mathbb{R}^{2n}}) \quad \text{as} \quad \|x\|_{\mathbb{R}^{2n}} \to \infty$$

uniformly in t.

We say that H is nonresonance at infinity, if

(10.10) $$\sigma(JB_\infty) \cap 2\pi i\mathbb{Z} = \emptyset.$$

LEMMA 10.1. *Suppose that* B *is a* $2n \times 2n$ *symmetric matrix. Let* F *be the linear operator defined by*

$$FZ(t) = BZ(t) \qquad \forall Z \in X.$$

Then $\sigma(JB) \cap 2\pi i\mathbb{Z} = \emptyset$ *is equivalent to*

(10.11) $$0 \notin \sigma(A-F).$$

PROOF. $\lambda \in \sigma(A-F) \iff \exists$ nontrivial 1-periodic $Z(t)$ such that

$$-J\dot{Z}(t) - BZ(t) = \lambda Z(t)$$

i.e.,

$$Z(t) = e^{t(JB+\lambda J)}Z(0)$$

or equivalently,

$$1 \in \sigma(e^{(JB+\lambda J)}) = e^{\sigma(JB+\lambda J)}$$

i.e.,

$$2\pi i\mathbb{Z} \cap \sigma(JB+\lambda J) \neq \emptyset.$$

Now let us define

$$F_\infty Z(t) = B_\infty Z(t)$$

then, by (10.9), we have

(10.12) $$\|F(Z) - F_\infty(Z)\|_X = o(\|Z\|_X) \quad \text{as} \quad \|Z\|_X \to \infty.$$

The finite dimensional reduction of Section 8 is employed. The problem (10.8) is reduced to a variational problem defined on $H_o = \bigoplus_{|j|\le C} M(j)$. Let P^o be the orthogonal projection onto H_o. We shall prove that $P^oF_\infty = F_\infty P^o$. In fact, $\forall j \in \mathbb{Z}$ the subspace $M(j) \oplus M(-j)$ is invariant under F_∞, which can be done by a simple computation.

In addition, we make some assumptions on the behaviour of H near θ:

(10.13) $$H_x(t,\theta) = \theta$$

(10.14) $$\|H_x(t,x) - B_o x\|_{\mathbb{R}^{2n}} = o(\|x\|_{\mathbb{R}^{2n}}) \quad \text{as} \quad \|x\|_{\mathbb{R}^{2n}} \to 0$$

where B_o is again a $2n \times 2n$ symmetric matrix. From which we have:

LEMMA 10.2. *Under the assumptions* (10.5), (10.13) *and* (10.14) *we have*

$$dF(\theta)Z = B_o Z(t) \qquad \forall Z \in X.$$

PROOF. $\forall Z_o \in X$, with $\|Z_o\| = 1$, let

$$E_M = \{t \in [0,1] \mid \|Z_o(t)\|_{\mathbb{R}^{2n}} > M\} \qquad \forall M \in \mathbb{R}^1.$$

Then one has

$$\text{mes}(E_M) \to 0 \quad \text{as} \quad M \to +\infty.$$

Since $\forall s > 0$

$$\|F(sZ_o) - sB_oZ_o\|_X^2 = \int_0^1 \|H_x(t,sZ_o(t)) - sB_oz_o(t)\|_{\mathbb{R}^{2n}}^2 \, dt$$
$$= \int_{E_M} + \int_{\complement E_M}$$

and by (10.5)

$$\|H_x(t,x)\|_{\mathbb{R}^{2n}} \le C\|x\|_{\mathbb{R}^{2n}},$$

we conclude

$$\int_{E_M} \|H_x(t,sZ_o(t)) - sB_oZ_o(t)\|^2_{\mathbb{R}^{2n}}\, dt \quad o(s^2) \quad \text{as} \quad M \to +\infty.$$

Fixing M and letting $s \to 0$, we have

$$\int_{\mathbf{C}E_M} \|H_x(t,sZ_o(t)) - sB_oZ_o(t)\|^2_{\mathbb{R}^{2n}}\, dt = o(s^2) \quad \text{as} \quad s \to 0,$$

by (10.14). The lemma is proved.

For a $2n \times 2n$ symmetric matrix B, let us denote

$$Q_j = 2\pi j \begin{pmatrix} 0 & -J \\ J & 0 \end{pmatrix} - \begin{pmatrix} B & 0 \\ 0 & B \end{pmatrix}, \qquad j = 1,2,\dots, \text{ and } Q_o = -B.$$

Let $m_j^{\pm}$ be the dimension of the positive/negative eigenspace of Q_j, $j = 0,1,\dots$

Depending on H, we define

$$m^{\pm}(B) = \sum_{j=0}^{[c]} m_j^{\pm}.$$

THEOREM 10.1. *Suppose that* $H \in C^2(\mathbb{R}^1\times\mathbb{R}^{2n},\mathbb{R}^1)$ *is* 1*-periodic with respect to* t, *and satisfies the conditions* (10.5), (10.9) *and* (10.10). *Then the Hamiltonian system* (10.1) *possesses at least one* 1*-periodic solution.*

Moreover, if (10.13) *and* (10.14) *are satisfied, and if either* $m^-(B_\infty) < m^-(B_o)$ *or* $m^+(B_\infty) < m^+(B_o)$, *then* (10.1) *possesses at least one nontrivial* 1*-periodic solution.*

PROOF. The first conclusion follows directly from Theorem 8.1 using Lemma 10.1. Now we turn to the second statement. Noticing that the finite dimensional reduction presented in Section 8 reduces our problem to finding critical points of a C^2-function a, defined on the finite dimensional space.

$$Z = H_o = \bigoplus\sum_{|j|\le[c/2\pi]} M(j),$$

and that each $M(j) \oplus M(-j)$, $j = 1,2,\dots,[c/2\pi]$ as well as $M(0)$, are invariant subspaces of both A and F_∞. Choosing a suitable basis $\{\cos 2\pi jte_k, \sin 2\pi jte_k \mid k = 1,\ldots,2n\}$ on $M(j) \oplus M(-j)$ ($\{e_k \mid k = 1,\ldots,2n\}$ on $M(0)$), the operator $A - F_\infty$ is represented by the following matrix

$$\begin{cases} 2\pi j\begin{pmatrix} 0 & -J \\ J & 0 \end{pmatrix} - \begin{pmatrix} B_\infty & 0 \\ 0 & B_\infty \end{pmatrix} & \text{on } M(j) \oplus M(-j) \text{ for } j > 0 \\ -B_\infty & \text{on } M(0). \end{cases}$$

It was proved in Section 8 that the asymptotics of $a'(z)$ at infinity is the matrix $A - F_\infty|_Z$, therefore its Morse index is counted to be $m^-(B_\infty)$. Similarly, the operator $d^2a(\theta) = A - dF(\theta)$ being represented by the matrix

$$\begin{cases} 2\pi j\begin{pmatrix} 0 & -J \\ J & 0 \end{pmatrix} - \begin{pmatrix} B_o & 0 \\ 0 & B_o \end{pmatrix} & \text{on } M(j) \oplus M(-j) \text{ for } j > 0 \\ -B_o & \text{on } M(0), \end{cases}$$

the Morse index of $d^2a(\theta)|_Z$ is counted to be $m^-(B_o)$. In either case, $m^+(B_\infty) < m^+(B_o)$ or $m^-(B_\infty) < m^-(B_o)$. On account of the fact (10.11), i.e., $A - F_\infty|_Z$ is invertible, we obtain

$$m^-(A - F_\infty|_Z) \notin [m^-(d^2a(\theta)), m^-(d^2a(\theta)) + \dim\ker(d^2a(\theta))].$$

The theorem is proved by Theorem 8.2.

COROLLARY 10.1. *Under the conditions in Theorem* 10.1, *if we assume further that*

$$\sigma(JB_o) \cap 2\pi i\mathbf{Z} = \emptyset,$$

and that the nontrivial solution obtained is nondegenerate, then the equation (10.1) *possesses at least two nontrivial* 1-*periodic solutions*.

Follows directly from Lemma 10.1 and Corollary 8.2.

Note: here we shall explain what is called a nondegenerate periodic solution of (10.1).

DEFINITION 10.1. Let $x(t)$ be a 1-periodic solution of (10.1). It is called *nondegenerate*, if there is no Floquet multiplier equal to 1 of the following linear periodic system:

$$-J\dot{w}(t) = \partial_x^2 H(t,x(t))w(t). \tag{10.15}$$

Let $t \mapsto B(t)$ be a continuous loop of symmetric matrices, with $B(t) = B(t+1)$. We say that λ is a *Floquet multiplier* of

$$-J\dot{w}(t) = B(t)w(t) \tag{10.16}$$

if λ is an eigenvalue of $\overline{W}(1)$, where $\overline{W}(t)$ is the fundamental solution matrix of (10.16). By definition, $x_o(t)$ is *nondegenerate* if and only if the system (10.15) has no nontrivial 1-periodic solution.

Let

$$a(z) = \tfrac{1}{2}(Au(z),u(z)) - \int_0^1 H(t,u(z))\,dt$$

be the function corresponding to the equation (10.2) obtained in Section 8, where $u(z) = z + \xi(z)$, $z \in H_o = Z$, and $\xi(z) \in Z^{\perp}$.

LEMMA 10.3. *Suppose that* u_o *is a nondegenerate* 1-*periodic solution of* (10.1). *Then the correspondence* $z_o \in Z$ *is a nondegenerate critical point of* a.

PROOF. By Lemma 10.2,

$$dF(u_o)z = \partial_x^2 H(t,u_o(t))z(t) \qquad \forall z \in X$$

and by definition, as well as the discreteness of the spectrum $\sigma(A - dF(u_o))$,

$$0 \notin \sigma(A - dF(u_o));$$

it follows that $(A - dF(u_o))^{-1}$ is bounded. However,

$$a''(z_o) = [A - dF(u_o)]u'(z_o),$$

where

$$u'(z_o) = id_Z + \xi'(z_o),$$

hence $a''(z_o)$ must be invertible, i.e., z_o is nondegenerate.

We shall extend this investigation to the case where the asymptotics B_o and B_∞ in Theorem 10.1 depend on t, i.e.,

$$H_x(t,x) = B_j(t)x + o(\|x\|_{\mathbb{R}^{2n}}) \quad \text{as} \quad \|x\| \to j, \quad j = 0,\infty,$$

uniformly in $t \in \mathbb{R}^1$, where $B_j(t)$ are continuous loops of symmetric matrices with $B_j(t) = B_j(t+1)$, $j = 0,\infty$.

Let P be the set of continuous loops with the additional property that the corresponding equation (10.16) has no Floquet multiplier equal to 1.

DEFINITION 10.2. Two loops $A_o(t),A_1(t) \in P$ are called *equivalent*, if one loop can be deformed into another without leaving the set P.

In order to distinguish distinct equivalence classes in P, Conley and Zehnder [30] defined an index for constant loops $A(t) = B$ such that JB has only simple purely imaginary eigenvalues, as follows.

If $\lambda \in \sigma(JB) \cap i\mathbb{R}^1$, then $\overline{\lambda} \in \sigma(JB) \cap i\mathbb{R}^1$. Let $(e,\overline{e})$ be the corresponding complex eigenvectors, then $\langle\overline{e},Je\rangle \neq 0$ is purely imaginary. We set

$$\alpha(\lambda) = \operatorname{sign}(-i\langle\overline{e},Je\rangle)\operatorname{Im}\lambda.$$

Observe that $\alpha(\lambda) = \alpha(\overline{\lambda})$ and $\alpha(\lambda) \notin 2\pi\mathbb{Z}$. There is an integer m such that $2m\pi < \alpha(\lambda) < 2(m+1)\pi$. We set

$$[\alpha(\lambda)] = m + \tfrac{1}{2}.$$

DEFINITION 10.3. For such a matrix B, we define

$$j(JB) = \sum_{\lambda\in\sigma(JB)\cap i\mathbb{R}^1} [\alpha(\lambda)]$$

to be the index of B.

Conley and Zehnder proved:

(THEOREM). *Each equivalence class of the set P of loops contains constant loops $A(t) = B$ for which $j(JA(t))$ is defined. All such constant loops in the same equivalence class have the same index, and constant loops in different components of P have different indices. To every integer $j \in \mathbf{Z}$ there is exactly one equivalence class having a constant loop with index j.*

This theorem enables us to extend the definition of index to loops in P.

According to Amann and Zehnder [6], one easily verifies: for all constant loops $B_0, B_\infty \in P$, we have

$$m^-(B_0) - m^-(B_\infty) = j(JB_0) - j(JB_\infty). \tag{10.17}$$

Theorem 10.1 then can be extended to be the following

THEOREM 10.2. *Suppose that there exist two loops $B_0(t), B_\infty(t) \in P$, with*

$$H_x(t,x) = B_j(t)x + o(\|x\|) \quad \text{as} \quad \|x\| \to j, \quad j = 0,\infty,$$

uniformly in $t \in \mathbb{R}^1$. Then the Hamiltonian system (10.2) *has at least one nontrivial 1-periodic solution provided $j(JB_0(t)) \neq j(JB_\infty(t))$.*

PROOF. By definition there exist two continuous families $\tilde{B}_\sigma(t)$ and $\hat{B}_\sigma(t)$ in P, $\sigma \in [0,1]$, such that

$$\tilde{B}_1(t) = B_0(t), \quad \tilde{B}_0(t) = B_0, \text{ a constant loop,}$$
$$\hat{B}_1(t) = B_\infty(t), \quad \hat{B}_0(t) = B_\infty, \text{ a constant loop.}$$

One then easily constructs a continuous family of Hamiltonian functions $H_\sigma(t,x)$, $\sigma \in [0,1]$, such that $H_1(t,x) = H(t,x)$ and that

$$H'_{\sigma x}(t,x) = \begin{cases} \tilde{B}_\sigma(t)x + \sigma(H'_x(t,x) - B_0(t)x) & \|x\| \le 1, \\ \hat{B}_\sigma(t)x + \sigma(H'_x(t,x) - B_\infty(t)x) & \|x\| \ge 2, \end{cases}$$

which implies that

$$H'_{0x}(t,x) = \begin{cases} B_0 & \|x\| < 1, \\ B_\infty & \|x\| \ge 2, \end{cases}$$

and

$$H'_{1x}(t,x) = H'_x(t,x).$$

Obviously, the condition (10.5) for the family $H_\sigma(t,x)$ holds, i.e., there is a constant $C > 0$ such that

$$\|\partial_x^2 H_\sigma(t,x)\| \le C.$$

This gives rise to the finite dimensional reduction of the whole family, i.e., there is a finite dimensional space Z which is invariant under the self-adjoint operator $A = -J\frac{d}{dt}$. The problem is thus reduced to finding critical points of the family of functions:

$$a_\sigma(z) = \tfrac{1}{2}(Ax(z),x(z)) - \Phi_\sigma(x(z)),$$

where

$$\Phi_\sigma(x) = \int_0^1 H_\sigma(t,x(t))\,dt.$$

According to (8.15),

$$a'_\sigma(z) = Ax(z) - F_\sigma(x(z)),$$

where $F_\sigma(x) = H'_{\sigma x}(t,x)$. Therefore

$$(10.18)\qquad a'_\sigma(z) = \begin{cases} (A - \tilde{B}_\sigma(t))x(z) - \sigma(F_1(x(z)) - B_0(t)x(z)) & \|x(z)\| \le 1 \\ (A - \hat{B}_\sigma(t))x(z) - \sigma(F_1(x(z)) - B_\infty(t)x(z)) & \|x(z)\| \ge 2. \end{cases}$$

Since $\hat{B}_\sigma$ is in P, by Lemma 10.3, $0 \notin \sigma(A - \hat{B}_\sigma)$, there exists a constant $\nu > 0$ independent of σ, such that

$$\|(A - \hat{B}_\sigma)x\| \ge \nu\|x\| \qquad \forall\sigma \in [0,1] \quad \forall x \in D(A)$$

where $\|.\|$ is the norm in $L^2((0,1),\mathbb{R}^{2n})$. By the asymptotic assumption we have

$$\|F_1(x) - B_\infty(t)x\| = o(\|x\|) \quad \text{as} \quad \|x\| \to \infty.$$

Therefore, there exists a constant $R > 0$ independent of σ such that

$$(10.19)\qquad \|a'(z)\| \ge \frac{\nu}{2}\|z\| \quad \text{as} \quad \|z\| \ge R.$$

Similarly, we conclude that there is an $\varepsilon > 0$, independent of σ, such that

$$\|a'_\sigma(z)\| \geq \frac{\nu}{2}\|z\| \quad \text{as} \quad \|z\| < \varepsilon. \tag{10.20}$$

(10.20) implies that θ is an isolated critical point of the whole family $a_\sigma(z)$. We apply Corollary 6.1 to conclude that the critical groups for a_σ at $z = \theta$ are invariant, i.e.,

$$C_q(a_\sigma,\theta) = C_q(a_o,\theta) = \begin{cases} G, & q = m^-(B_o), \\ 0, & q \neq m^-(B_o). \end{cases} \tag{10.21}$$

The conclusion (10.19) implies that all the critical sets $K(\sigma)$ of the family of functions a_σ, $\sigma \in [0,1]$, are contained in an open ball, say B_R. Since the $|a_\sigma|$ are uniformly bounded by a constant M on B_R, we may choose $\mathcal{O} = B_R$, and a bounded set N, depending on R, M and ν, such that Theorem 6.4 is applicable. From which we obtain

$$H_*((a_1)_M \cap B_R,((a_1)_M\backslash[K(1)]) \cap B_R) \cong H_*((a_o)_M \cap B_R,((a_o)_M\backslash[K(0)]) \cap B_R).$$

However, if we split the space Z according to the positive and negative eigenspace of the linear operator $A - B_\infty$: $Z = Z_+ \oplus Z_-$, we have

$$\begin{aligned} &H_q((a_o)_M \cap B_{R_1},((a_o)_M\backslash[K(0)]) \cap B_{R_1}) \\ &\cong H_q((a_o)_M \cap ((Z_+\cap B_{R_1}) \times Z_-),((a_o)_M\backslash[K(0)]) \cap ((Z_+\cap B_{R_1}) \times Z_-)) \\ &\cong H_q((a_o)_M \cap ((Z_+\cap B_{R_1}) \times Z_-),((a_o)_{-M} \cap (Z_+\cap B_{R_1}) \times Z_-)) \\ &\cong \begin{cases} G, & q = m^-(B_\infty) \\ 0, & q \neq m^-(B_\infty) \end{cases} \end{aligned}$$

for suitable $R_1 > R$, provided by Theorem 8.2.

Since $j(JB_o(t)) \neq j(JB_\infty(t))$, we have $j(JB_o) \neq j(JB_\infty)$, which implies $m^-(B_o) \neq m^-(B_\infty)$ by (10.17).

The conclusion then follows from the fact that

$$H_q((a_1)_M \cap B_R, ((a_1)_M \setminus [K(1)]) \cap B_R) \cong H_q((a_1)_M, (a_1)_{-M}), \qquad q = 0,1,2,\ldots$$

If there were only one critical point of a_1, this would contradict (10.21). The proof is finished.

REMARK 10.1. Theorems 10.1, 10.2 are based on Amann and Zehnder [5], Chang [21] and Conley and Zehnder [31], but avoiding the use of Conley's Morse theory.

10.2. A multiple solution theorem

Now we turn to a result which is concerned with the existence of at least two nontrivial periodic solutions.

THEOREM 10.3. *Suppose that* $H \in C^2(\mathbb{R}^1 \times \mathbb{R}^{2n}, \mathbb{R}^1)$ *satisfies the following conditions:*

(1) *There exist constants* $\alpha < \beta$ *such that*

$$\alpha I \le \partial_x^2 H(t,x) \le \beta I \qquad \forall (t,x) \in \mathbb{R}^1 \times \mathbb{R}^{2n}.$$

(2) *Let* $j_0, j_0+1, \ldots, j_1$ *be all integers within* $[\frac{\alpha}{2\pi}, \frac{\beta}{2\pi}]$ *(without loss of generality, we may assume* $\alpha, \beta \notin 2\pi\mathbb{Z}$*). Suppose that there exist* γ *and* C*, such that* $2\pi j_1 < \gamma < \beta$ *and*

$$H(t,x) \ge \tfrac{1}{2}\gamma \|x\|^2 - C \qquad \forall (t,x) \in \mathbb{R}^1 \times \mathbb{R}^{2n}.$$

(3) $H_x(t,\theta) = \theta$. $\exists j \in [j_0, j_1) \cap \mathbb{Z}$ *such that*

$$2\pi j I < \partial_x^2 H(t,\theta) < 2\pi(j+1) I \qquad \forall t \in \mathbb{R}^1.$$

(4) H *is* 1*-periodic with respect to* t.

Then the Hamiltonian system (10.1) *possesses at least two nontrivial periodic solutions.*

PROOF. According to the finite dimensional reduction we turn to the function

$$a(z) = \tfrac{1}{2}(Au(z),u(z)) - \int_0^1 H(t,u(z))\,dt,$$

where $u(z) = z + u_+(z) + u_-(z)$, $z \in Z \overset{\Delta}{=} H_0$, and $u_\pm(z) \in H_\pm$. Since

$$a(z) = \tfrac{1}{2}(Aw,w) - \int_0^1 H(t,w)\,dt$$

$$+ \{\tfrac{1}{2}[(Au(z),u(z)) - (Aw,w)] - \int_0^1 [H(t,u(z)) - H(t,w)]\,dt\},$$

where $w = z + u_-(z)$, and the terms in the bracket are equal to

$$-\int_0^1 (F(su_+ + w),u_+)\,ds + \tfrac{1}{2}(Au(z),u_+)$$

$$= \int_0^1 \int_0^1 (\partial_x^2 H(t,su_+ + w)u_+,u_+)s\,ds\,dt - \tfrac{1}{2}(Au_+,u_+) \le 0$$

by condition (1), we obtain

$$a(z) \le \tfrac{1}{2}(Aw,w) - \int_0^1 H(t,w)\,dt$$

$$\le \tfrac{1}{2}(2\pi j_1 - \gamma)\|w\|^2 + C \to -\infty \quad \text{as} \quad \|z\| \to \infty$$

using condition (2).

Therefore the function $-a(z)$ is bounded from below and satisfies the P.S. condition.

In order to apply the three critical point theorem studied in Section 5, we claim that θ is neither a minimum, nor degenerate. In fact, using condition (3), it follows from Lemma 8.3, that

$$\tfrac{1}{2}((A - \bar{\lambda}I)z,z) + o(\|z\|^2) \le a(z) \le \tfrac{1}{2}((A - \hat{\lambda}I)z,z) + o(\|z\|^2)$$

where $(\bar{\lambda},\hat{\lambda}) \subset 2\pi(j,j+1)$, as $\|z\| \to 0$. The theorem is proved.

REMARK 10.2. An abstract version of this theorem can be found in Chang [21], which includes the counterpart in semilinear elliptic BVP as special case, cf. Castro and Lazer [20].

10.3. Arnold's conjecture

Arnold conjectured that every symplectic diffeomorphism of a 2-dimensional torus T^2 which is homologous to the identity, has at least three distinct fixed points, cf. [11], [12].

The fixed point problem is equivalently reduced to finding 1-periodic solutions of the Hamiltonian system:

$$\dot{x} = JH_x(t,x),$$

where $H \in C^2(\mathbb{R}^1 \times \mathbb{R}^{2n}, \mathbb{R}^1)$ is 1-periodic with respect to each variable [30].

Now we shall prove:

THEOREM 10.4. *Suppose* $H \in C^2(\mathbb{R}^1 \times \mathbb{R}^{2n}, \mathbb{R}^1)$ *is* 1-*periodic with respect to each variable, then there are at least* $2n+1$ *distinct* 1-*periodic solutions of the system* (10.1). *If further we assume that all* 1-*periodic solutions are nondegenerate, then* (10.1) *possesses at least* 2^{2n} *distinct* 1-*periodic solutions.*

PROOF. We have seen that the problem is reduced to finding critical points of the following finite dimensional function:

$$a(z) = \tfrac{1}{2}(Au(z),u(z)) - \int_0^1 H(t,u(z))\,dt,$$

where $z \in Z = H_o \overset{\Delta}{=} \bigoplus\sum_{|j| \le [c]/2\pi} M(j)$, $u(z) = z + \xi(z)$, $\xi(z) \in Z^{\perp}$. Decomposing the space Z again:

$$Z = Z_+ \oplus Z_o \oplus Z_-,$$

where

$$Z_+ = \bigoplus\sum_{0<j\le[c]/2\pi} M(j), \quad Z_o = M(0) \quad \text{and} \quad Z_- = \bigoplus\sum_{-[c]/2\pi \le j<0} M(j),$$

we write

$$z = z_+ + z_o + z_-, \quad \text{where} \quad z_\pm \in Z_\pm \quad \text{and} \quad z_o \in Z_o.$$

Since $Fx = H_x(t,z+\xi(z))$, which is 1-periodic with respect to z_o, the unique solution $\xi(z)$, which solves the equation (8.11), must be 1-periodic with respect to z_o also. Therefore $a(z)$ is a 1-periodic function with respect to z_o.

Now the function a is regarded to be a function defined on $Z_+ \times Z_- \times T^{2n}$. Noticing that

(1) $Az_o = \theta$,

(2) $\|\xi(z)\|$ is bounded,

we denote $u = (z_+,z_-)$, and

$$g(u,z_o) = \tfrac{1}{2}(A\xi(z),\xi(z)) - \int_0^1 H(t,z+\xi(z))\, dt;$$

then

$$a(z) = \tfrac{1}{2}(Au,u) + g(u,z_o),$$

where $g \in C^2(Z_+ \times Z_- \times T^{2n};\mathbb{R}^1)$, having bounded compact gradient vector field. Since

$$\text{cuplength}(T^{2n}) = 2n,$$

we obtain at least $2n+1$ distinct critical points of a applying Theorem 8.3. As to the second conclusion, we know that

$$\text{rank } H_k(T^{2n}) = C_k^{2n}, \qquad k = 0,1,2,\dots,2n;$$

is combining with Lemma 10.3, the proof follows directly via Theorem 8.3.

REMARK 10.3. When $n = 2$, this is just the answer to Arnold's conjecture. However, the first proof of this theorem was given by Conley and Zehnder [30] via Conley's generalization of the Morse index. There are many different proofs, cf. Weinstein [84], Hofer, Berestycki and Lasry. The above proof is taken from Chang [25].

Now we turn to point out some other interesting results on Hamiltonian systems which can be obtained by Morse theory. If the Hamiltonian function H is independent of time t, there are two types of periodic solution problems: (1) prescribed period; this has been discussed above; (2) prescribed energy. This is well known from the fact that if $x(t)$ is a solution of (10.2), then

$$\begin{aligned}\frac{d}{dt} H(x(t)) &= (H'(x(t)),\dot{x}(t)) \\ &= (H'(x(t)),JH'(x(t)) = 0.\end{aligned}$$

Hence, introducing the energy surface

$$M_c = \{x \in \mathbb{R}^{2n} \mid H(x) = c\}$$

for some $c \in R$, we are looking for periodic solutions of (10.2) with $x(t) \in M_c$.

One asks: if $M_c = \partial C$ is the C^2 boundary of a bounded and strictly convex domain $C \subset \mathbb{R}^{2n}$, how many periodic solutions of (10.2) are there on the surface M_c?

P.H. Rabinowitz [68] and A. Weinstein [83] proved that in case

$$H'(x) \neq \theta \qquad \forall x \in M_c \tag{10.22}$$

there is at least one periodic solution on M_c.

Ekeland and Lasry [34] proved that in addition to the assumption (10.22), if $D_r \subset C \subset D_R$, for two disks with radii $r < R < \sqrt{2}\, r$, then M_c carries at least n distinct periodic orbits.

Recently, Ekeland [33] obtained the following interesting result via Morse theory: If $n \geq 3$, then for almost all convex Hamiltonian systems (10.2), there are infinitely many distinct periodic orbits on the surface M_c.

At the end, I would like to mention a few references on the applications of Morse theory to estimates of the number of minimal surfaces: Morse and Thompkins [58], Struwe [78] and Tromba [80].

REFERENCES

[1] AMANN, H., Fixed point equations and nonlinear eigenvalue problems in ordered Banach spaces, *SIAM Rev.* 18(1976), 620-709.

[2] AMANN, H., Saddle points and multiple solutions of differential equations, *Math. Z.* 169(1979), 127-166.

[3] AMANN, H., A note on degree theory for gradient mappings, *Proc. Amer. Math. Soc.* 85(1982), 591-597.

[4] AMANN, H., HESS, P., A multiplicity result for a class of elliptic boundary value problems, *Proc. Roy. Soc. Edinburgh Sect. A* 84(1979), 145-151.

[5] AMANN, H., ZEHNDER, E., Nontrivial solutions for a class of non-resonance problems and applications to nonlinear differential equations, *Annali Scuola Norm. Sup. Pisa Cl. Sci.* (4) 7(1980), 539-603.

[6] AMANN, H., ZEHNDER, E., Periodic solutions of asymptotically linear Hamiltonian systems, *Manuscripta Math.* 32(1980), 149-189.

[7] AMBROSETTI, A., Differential equations with multiple solutions and nonlinear functional analysis, ISAS, Trieste, 1982, preprint.

[8] AMBROSETTI, A., Elliptic equations with jumping nonlinearities, Ecole polytechnique fédérale de Lausanne, 1982, preprint.

[9] AMBROSETTI, A., PRODI, G., On the inversion of some differentiable mappings with singularities between Banach spaces, *Ann. Mat. Pura Appl.* (4) 93(1973), 231-247.

[10] AMBROSETTI, A., RABINOWITZ, P.H., Dual variational methods in critical point theory and applications, *J. Funct. Anal.* 14(1973), 349-381.

[11] ARNOLD, V.I., Problems in present day mathematics, XX: Fixed points of symplectic diffeomorphisms, *Mathematical Developments Arising from Hilbert Problems* (F.E.Browder, ed.), Proc. Symp. Pure Math. vol. 28, American Mathematical Society, Providence 1976, p. 66.

[12] ARNOLD, V.I., *Mathematical Methods of Classical Mechanics*, Springer, Berlin-Heidelberg-New York, 1978.

[13] ATIYAH, M., BOTT, R., On the Yang-Mills equations over Riemann surfaces, 1983, preprint.

[14] BERESTYCKI, H., LIONS, P.L., Sharp existence results for a class of semi-linear elliptic problems, *Bol. Soc. Brasil. Mat.* 12(1981), 9-20.

[15] BERGER, M.S., *Nonlinearity and Functional Analysis*, Academic Press, New York, 1977.

[16] BERGER, M.S., PODOLAK, E., On the solutions of a nonlinear Dirichlet problem , *Indiana Univ. Math. J.* 24(1975), 837-846.

[17] BONIC, R., FRAMPTON, J., Smooth functions on Banach manifolds, *J. Math. and Mech.* 15(1966), 877-898.

[18] BOTT, R., Nondegenerate critical manifolds, *Ann. of Math.* 60(1954), 248-261.

[19] BOTT, R., Lectures on Morse theory, old and new, *Bull. Amer. Math. Soc.* 7(1982), 331-358.

[20] CASTRO, A., LAZER, A.C., Critical point theory and the number of solutions of a nonlinear Dirichlet problem, *Ann. Mat. Pura Appl.* (4) 70(1979), 113-137.

[21] CHANG, K.C., Solutions of asymptotically linear operator equations via Morse theory, *Comm. Pure Appl. Math.* 34(1981), 693-712.

[22] CHANG, K.C., Morse theory on Banach spaces and its applications, *Chinese Ann. Math. Ser. B* 4(1983), 381-399.

[23] CHANG, K.C., A variant mountain pass lemma, *Sci. Sinica Ser. A* 26(1983), 1241-1255.

[24] CHANG, K.C., Variational method and the sub- and super-solutions, *Sci. Sinica Ser. A* 26(1983), 1256-1265.

[25] CHANG, K.C., Applications of homology theory to some problems in differential equations, MSRI 064-83 (1983), Berkeley. To appear in *Nonlinear Functional Analysis and Applications* (F.E.Browder, ed.), Proc. Symp. Pure Math., American Mathematical Society, Providence.

[26] CHANG, K.C., On a bifurcation theorem due to Rabinowitz, *J. Systems Sci. Math. Sci.* 4(1984), 191-195.

[27] CHANG, K.C., WU, S.P., LI, S., Multiple periodic solutions for an asymptotically linear wave equation, *Indiana Math. J.* 31(1982), 721-731.

[28] CLARK, D.C., A variant of Ljusternik-Schnirelman theory, *Indiana Math. J.* 22(1972), 65-74.

[29] CONLEY, C.C., *Isolated Invariant Sets and the Morse Index*, CBMS Regional Conference Series 38, American Mathematical Society, Providence, 1978.

[30] CONLEY, C.C., ZEHNDER, E., The Birkhoff-Lewis fixed point theorem and a conjecture of V. Arnold, *Invent. Math.* 73(1983), 33-49.

[31] CONLEY, C.C., ZEHNDER, E., Morse type index theory for flows and periodic solutions for Hamiltonian equations, *Comm. Pure Appl. Math.* 37(1984), 207-253.

[32] DANCER, E.N., Degenerate critical points, homotopy indices and Morse inequalities, 1983, preprint.

[33] EKELAND, I., Une théorie de Morse pour les systèmes hamiltoniens convexes, *Ann. Inst. H. Poincaré, Anal. Non Linéaire* 1(1984), 19-78.

[34] EKELAND, I., LASRY, J.M., On the number of closed trajectories for a Hamiltonian flow on a convex energy surface, *Ann. of Math.* 112(1980), 283-319.

[35] FADELL, E.R., RABINOWITZ, P.H., Bifurcation for odd potential operators and an alternative topological index, *J. Funct. Anal.* 26(1977), 48-67.

[36] FUČIK, S., Remarks on a result by A. Ambrosetti and G. Prodi, *Boll. Un. Mat. Ital.* 11(1975), 259-267.

[37] GILBARG, D., TRUDINGER, N.S., *Elliptic Partial Differential Equations of Second Order*, Springer, Berlin-Heidelberg-New York, 1977.

[38] GREENBERG, M.J., *Lectures on Algebraic Topology*, Benjamin, New York, 1967.

[39] GROMOLL, D., MEYER, W., On differentiable functions with isolated critical points, *Topology* 8(1969), 361-369.

[40] HESS, P., On a nonlinear elliptic boundary value problem of the Ambrosetti-Prodi type, *Boll. Un. Mat. Ital.* (5), 17-A(1980), 187-192.

[41] HESS, P., KATO, T., On some linear and nonlinear eigenvalue problems with indefinite weight functions, *Comm. Partial Differential Equations* 5(1980), 999-1030.

[42] HOFER, A., Variational and topological methods in partially ordered Hilbert spaces; *Math. Ann.* 261(1982), 493-514.

[43] HOFER, H., A note on the topological degree at a critical point of mountainpass-type, *Proc. Amer. Math. Soc.* 90(1984), 309-315.

[44] KAZDAN, J.L., WARNER, F.W., Remarks on some quasilinear elliptic equations, *Comm. Pure Appl. Math.* 28(1975), 567-597.

[45] KLINGENBERG, W., *Lectures on Closed Geodesics*, Springer, Berlin-Heidelberg-New York, 1978.

[46] KRASNOSEL'SKII, M.A., *Topological Methods in the Theory of Nonlinear Integral Equations*, Pergamon, New York, 1964.

[47] KRASNOSEL'SKII, M.A., *The Operator of Translation Along the Trajectories of Differential Equations*, Translations of Mathematical Monographs, Vol. 19, American Mathematical Society, Providence, 1968.

[48] KUIPER, N., C^1-equivalence of functions near isolated critical points, *Symposium on Infinite-Dimensional Topology*, Annals of Mathematics Studies 69, Princeton University Press, Princeton, 1972.

[49] LAZER, A.C., McKENNA, P.J., On the number of solutions of a nonlinear Dirichlet problem, *J. Math. Anal. Appl.* 84(1981), 282-294.

[50] LIONS, J.-L., MAGENES, E., *Nonhomogeneous Boundary Value Problems and Applications*, Springer, Berlin-Heidelberg-New York, 1972.

[51] LIU, G.Q., Doctoral thesis, System Science Institute, Academy of Science, Beijing, 1983.

[52] LLOYD, N.G., *Degree Theory*, Cambridge University Press, Cambridge, 1977.

[53] MARINO, A., PRODI, G., La teoria di Morse per spazi di Hilbert, *Rend. Sem. Mat. Univ. Padova* 41(1968), 43-68.

[54] MARINO, A., PRODI, G., Metodi perturbativi nella teoria di Morse, *Boll. Un. Math. Ital. Suppl.* Fasc. 3(1975), 1-32.

[55] MILNOR, J., *Morse Theory*, Princeton University Press, Princeton, 1963.

[56] MILNOR, J., *Topology from the Differentiable Viewpoint*, University Press of Virginia, Charlottesville, 1969.

[57] MORSE, M., CAIRNS, S.S., *Critical Point Theory in Global Analysis and Differential Topology*, Academic Press, New York, 1969.

[58] MORSE, M., TOMPKINS, C., The existence of minimal surfaces of general critical types, *Ann. of Math.* (2) 40(1939), 443-472.

[59] NIRENBERG, L., *Topics in Nonlinear Functional Analysis*, Courant Institute Lecture Notes, New York, 1974.

[60] NIRENBERG, L., Variational and topological methods in nonlinear problems, *Bull. Amer. Math. Soc.* 4(1981), 267-302.

[61] PALAIS, R.S., Morse theory on Hilbert manifolds, *Topology* 2(1963), 299-340.

[62] PALAIS, R.S., Homotopy theory of infinite dimensional manifolds, *Topology* 5(1966), 1-16.

[63] PALAIS, R.S., Lusternik-Schnirelman theory on Banach manifolds, *Topology* 5(1966), 115-132.

[64] PALAIS, R.S., Critical point theory and the minimax principle, *Global Analysis*, Proc. Symp. Pure Math. 15 (ed. S.S. Chern), Amer. Math. Soc., Providence, 1970, 185-202.

[65] PALAIS, R.S., SMALE, S., A generalized Morse theory, *Bull. Amer. Math. Soc.* 70(1964), 165-171.

[66] RABINOWITZ, P.H., Variational methods for nonlinear eigenvalue problems, *Eigenvalues of Nonlinear Problems*, Edizioni Cremonese, Roma, 1974, 141-195.

[67] RABINOWITZ, P.H., A bifurcation theorem for potential operators, *J. Funct. Anal.* 25(1977), 412-424.

[68] RABINOWITZ, P.H., Periodic solutions of Hamiltonian systems, *Comm. Pure Appl. Math.* 31(1978), 157-184.

[69] RIESZ, F., ST-NAGY, B., *Leçons d'analyse fonctionnelle*, Akadémiai Kiadó, Budapest, 1955.

[70] ROTHE, E., Critical point theory in Hilbert space under regular boundary conditions, *J. Math. Anal. Appl.* 36(1971), 377-431.

[71] ROTHE, E., Morse theory in Hilbert space, *Rocky Mountain J. Math.* 3(1973), 251-274.

[72] ROTHE, E., On the connection between critical point theory and Leray Schauder degree, *J. Math. Anal. Appl.* 88(1982), 265-269.

[73] RYBAKOWSKI, K., On homotopy index for infinite-dimensional semiflows, *Trans. Amer. Math. Soc.* 269(1982), 351-381.

[74] SMALE, S., Differentiable dynamical systems, *Bull. Amer. Math. Soc.* 70 (1967), 747-817.

[75] SCHWARTZ, J.T., *Nonlinear Functional Analysis*, Gordon Breach, New York, 1969.

[76] SOLIMINI, S., Existence of a third solution for a class of BVP with jumping nonlinearities, Preprint.

[77] STRUWE, M., A note on a result of Ambrosetti and Mancini, Preprint no. 496, Universität Bonn, 1982.

[78] STRUWE, M., On a critical point theory for minimal surfaces spanning a wire in $\mathbb{R}^n$, *J. Reine Angew. Math.* 349(1984), 1-23.

[79] TIAN, G., On the mountain pass theorem, *Kexue Tongbao* 29(1984), 1150-1154.

[80] TROMBA, A.J., *On the Number of Simply Connected Minimal Surfaces Spanning a Curve*, Memoirs AMS no. 194, American Mathematical Society, Providence, 1977.

[81] TROMBA, A.J., A general approach to Morse theory, *J. Differential Geom.* 12(1977), 47-85.

[82] UHLENBECK, K., Morse theory on Banach manifolds, *J. Funct. Anal.* 10(1972), 430-445.

[83] WEINSTEIN, A., Periodic orbits for convex Hamiltonian systems, *Ann. of Math.* 108(1978), 507-518.

[84] WEINSTEIN, A., C^0-perturbation theorems for symplectic fixed points and Lagrangian intersections, Preprint.

LES PRESSES DE L'UNIVERSITÉ DE MONTRÉAL
C.P. 6128, succc. «A», Qué., Canada H3C 3J7

EXTRAIT DU CATALOGUE

Mathématiques

COLLECTION «SÉMINAIRE DE MATHÉMATIQUES SUPÉRIEURES»

1. **Problèmes aux limites dans les équations aux dérivées partielles.** Jacques L. LIONS
2. **Théorie des algèbres de Banach et des algèbres localement convexes.** Lucien WAELBROECK
3. **Introduction à l'algèbre homologique.** Jean-Marie MARANDA
4. **Séries de Fourier aléatoires.** Jean-Pierre KAHANE
5. **Quelques aspects de la théorie des entiers algébriques.** Charles PISOT
6. **Théorie des modèles en logique mathématique.** Aubert DAIGNEAULT
7. **Promenades aléatoires et mouvements brownien.** Anatole JOFFE
8. **Fondements de la géométrie algébrique moderne.** Jean DIEUDONNÉ
9. **Théorie des valuations.** Paulo RIBENBOIM
10. **Catégories non abéliennes.** Peter HILTON, Tudor GANEA, Heinrich KLEISLI, Jean-Marie MARANDA, Howard OSBORN
11. **Homotopie et cohomologie.** Beno ECKMANN
12. **Intégration dans les groupes topologiques.** Geoffrey FOX
13. **Unicité et convexité dans les problèmes différentiels.** Shmuel AGMON
14. **Axiomatique des fonctions harmoniques.** Marcel BRELOT
15. **Problèmes non linéaires.** Félix E. BROWDER
16. **Équations elliptiques du second ordre à coefficients discontinus.** Guido STAMPACCHIA
17. **Problèmes aux limites non homogènes.** José BARROS-NETO
18. **Équations différentielles abstraites.** Samuel ZAIDMAN
19. **Équations aux dérivées partielles.** Robert CARROL, George F.D. DUFF, Jöran FRIBERG, Jules GOBERT, Pierre GRISVARD, Jindrich NECAS et Robert SEELEY
20. **L'Algèbre logique et ses rapports avec la théorie des relations.** Roland FRAÏSSÉ
21. **Logical Systems Containing Only a Finite Number of Symbols.** Leon HENKIN
24. **Représentabilité et définissabilité dans les algèbres transformationnelles et dans les algèbres polyadiques.** Léon LEBLANC
25. **Modèles transitifs de la théorie des ensembles de Zermelo-Fraenkel.** Andrzej MOSTOWSKI
26. **Théorie de l'approximation des fonctions d'une variable complexe.** Wolfgang H.J. FUCHS
27. **Les Fonctions multivalentes.** Walter K. HAYMAN
28. **Fonctionnelles analytiques et fonctions entières (n variables).** Pierre LELONG
29. **Applications of Functional Analysis to Extremal Problems for Polynomials.** Qazi Ibadur RAHMAN
30. **Topics in Complex Manifolds.** Hugo ROSSI
31. **Théorie de l'inférence statistique robuste.** Peter J. HUBER
32. **Aspects probabilistes de la théorie du potentiel.** Mark KAC
33. **Théorie asymptotique de la décision statistique.** Lucien M. LECAM
34. **Processus aléatoires gaussiens.** Jacques NEVEU
35. **Nonparametric Estimation.** Constance van EEDEN
36. **K-Théorie.** Max KAROUBI
37. **Differential Complexes.** Joseph J. KOHN
38. **Variétés hilbertiennes : aspects géométriques.** Nicolaas H. KUIPER
39. **Deformations of Compact Complex Manifold.** Masatake KURANISHI
40. **Grauert's Theorem of Direct Images of Coherent Sheaves.** Raghavan NARASIMHAN
41. **Systems of Linear Partial Differential Equations and Deformation of Pseudogroup Structures.** A. KUMPERA et D.C. SPENCER
42. **Analyse globale.** P. LIBERMANN, K.D. ELWORTHY, N. MOULIS, K.K. MUKHERJEA, N. PRAKASH, G. LUSZTIC et W. SHIH
43. **Algebraic Space Curves.** Sheeram S. ABHYANKAR
44. **Théorèmes de représentabilité pour les espaces algébriques.** Michael ARTIN
45. **Groupes de Barsotti-Tate et cristaux de Dieudonné.** Alexandre GROTHENDIECK
46. **On Flat Extensions of a Ring.** Massayoshi NAGATA
47. **Introduction à la théorie des sites et son application à la construction des préschémas quotients.** Masayoshi MIYANISHI
48. **Méthodes logiques en géométrie diophantienne.** Shuichi TAKAHASHI
49. **Index Theorems of Atiyah — Bott — Patodi and Curvature Invariants.** Ravindra S. KULKARNI
50. **Numerical Methods in Statistical Hydrodynamics.** Alexandre CHORIN
51. **Introduction à la théorie des hypergraphes.** Claude BERGE
52. **Automath, a Language for Mathematics.** Nicolaas G. DE BRUIJN
53. **Logique des topos (Introduction à la théorie des topos élémentaires).** Dana SCHLOMIUK
54. **La Série génératrice exponentielle dans les problèmes d'énumération.** Dominique FOATA
55. **Feuilletages : résultats anciens et nouveaux (Painlevé, Hector et Martinet).** Georges H. REEB

56. **Finite Embedding Theorems for Partial Designs and Algebras.** Charles C. LINDNER et Trevor EVANS
57. **Minimal Varieties in Real and Complex Geometry**, H. Blaine LAWSON, J[r]
58. **La Théorie des points fixes et ses applications à l'analyse.** Kazimierz GEBA, Karol BORSUK, Andrzej JANKOWSKI et Edward ZHENDER
59. **Numerical Analysis of the Finite Element Method**, Philippe G. CIARLET
60. **Méthodes numériques en mathématiques appliquées.** J.F. Giles AUCHMUTY, Michel CROUZEIX, Pierre JAMET, Colette LEBAUD, Pierre LESAINT et Bertrand MERCIER
61. **Analyse numérique matricielle.** Paul ARMINJON
62. **Problèmes d'optimisation en calcul des probabilités.** Serge DUBUC
63. **Chaînes de Markov sur les permutations.** Gérard LETAC
64. **Géométrie différentielle stochastique.** Paul MALLIAVIN
65. **Numerical Methods for Solving Time-Dependent Problems for Partial Differential Equations.** Heinz-Otto KREISS
66. **Difference Sets in Elementary Abelian Groups.** Paul CAMION
67. **Groups in Physics : Collective Model of the Nucleus; Canonical Transformation in Quantum Mechanics.** Marcos MOSHINSKY
68. **Points fixes pour les applications compactes : espaces de Lefschetz et la théorie de l'indice.** Andrzej GRANAS
69. **Set Theoretic Methods in Homological Algebra and Abelian Groups.** Paul EKLOF
70. **Abelian p-Groups and Mixed Groups.** Laszlo FUCHS
71. **Integral Representations and Structure of Finite Group Rings.** Klaus W. ROGGENKAMP
72. **Homological Invariants of Modules over Commutative Rings.** Paul ROBERTS
73. **Representations of Valued Graphs.** Vlastimil DLAB
74. **Groupes abéliens sans torsion.** Khalid BENABDALLAH
75. **Lie Groups, Lie Algebras and Representation Theory.** Hans ZASSENHAUS
76. **Birational Geometry for Open Varieties.** Shigeru IITAKA
77. **Lectures on Hilbert Modular Surfaces.** Friedrich HIRZEBRUCH, Gerard van der GEER
78. **Complex Geometry in Mathematical Physics.** R.O. WELLS, Jr.
79. **Lectures on Approximation and Value Distribution.** Tord GANELIUS, Walter K. KAYMAN, Donald J. NEWMAN
80. **Sur la topologie des surfaces complexes compactes.** Srinivasacharyulu KILAMBI, Gottfried BARTHEL, Ludger KAUP
81. **Topics in Polynomial and Rational Interpolation and Approximation.** Richard S. VARGA
82. **Approximation uniforme qualitative sur des ensembles non bornés.** Paul M. GAUTHIER, Walter HENGARTNER
83. **Convolutions in Geometric Function Theory.** Stephan RUSCHEWEYH.
84. **Characteristic Properties of Quasidisks.** Frederick W. GEHRING.
85. **Curves in Projective Space.** Joe HARRIS.
86. **Les Inégalités de Markoff et de Bernstein.** Qazi Ibadur RAHMAN, Gerhard SCHMEISSER
87. **Analyse des données.** Francis CAILLIEZ
88. **Inférence statistique et analyse des données sous des plans d'échantillonnages complexes.** Carl E. SÃRNDAL
89. **Analyse de données chronologiques.** Guy MÉLARD
90. **Schauder's Estimates and Boundary Value Problems for Quasilinear Partial Differential Equations.** Manfred KÖNIG
91. **Topological Methods in Bifurcation Theory.** Kazimierz GĘBA et Paul H. RABINOWITZ
92. **Points fixes, points critiques et problèmes aux limites.** Jean MAWHIN
93. **Problèmes aux limites non linéaires pour certaines classes d'équations différentielles ordinaires.** Ronald B. GUENTHER
94. **The Fixed Point Index and Some Applications.** Roger D. NUSSBAUM
95. **Méthodes topologiques en analyse non linéaire.** Comptes rendus colligés par Andrzej Granas
96. **Analysis of Categorical Data.** Gary G. Koch, Peter B. Imrey *et al.*

COLLECTION « CHAIRE AISENSTADT »

Physical Aspects of Lie Group Theory. Robert HERMANN
Quelques problèmes mathématiques en physique statistique. Mark KAC
La Transformation de Weyl et la fonction de Wigner : une forme alternative de la mécanique quantique. Sybren DE GROOT
Sur quelques questions d'analyse, de mécanique et de contrôle optimal. Jacques Louis LIONS
Mariages stables et leurs relations avec d'autres problèmes combinatoires. Donald E. KNUTH
Symétries, jauges et variétés de groupe. Yuval NE'EMAN
La Théorie des sous-gradients et ses applications à l'optimisation. Fonctions convexes et non convexes. R. Tyrrel ROCKAFELLAR

Achevé d'imprimer
en janvier 1986 sur les presses
des Ateliers Graphiques Marc Veilleux Inc.
Cap-Saint-Ignace, Qué.